Mathematical Marvels

A Primer on

LOGARITHMS

Shailesh Shirali

Universities Press

Universities Press (India) Private Limited

Registered Office
3-6-747/1/A & 3-6-754/1, Himayatnagar,
Hyderabad 500 029 (A.P.), India
e-mail: info@universitiespress.com

Distributed by
Orient Blackswan Private Limited

Registered Office
3-6-752, Himayatnagar, Hyderabad 500 029 (A.P.), India

Other Offices
Bangalore / Bhopal / Bhubaneshwar / Chennai / Ernakulam
Guwahati / Hyderabad / Jaipur / Kolkata / Lucknow / Mumbai
New Delhi / Patna

© Universities Pres (India) Private Limited 2002

First published 2002
Reprinted 2010

ISBN: 978 81 7371 414 6

Typeset by
OSDATA, Hyderabad 500 029

Printed in India at
Sri Kalanjali Graphics
Hyderbad

Published by
Universities Press (India) Private Limited
3-6-747/1/A & 3-6-754/1, Himayatnagar
Hyderabad 500 029 (A.P), India

Contents

Preface

This book, addressed to school students, is about logarithms and the rich mathematical content they possess.

In past times, logarithms as a topic in the school curriculum was essentially about computation. The use of common logarithms was taught at the 9th and 10th standard levels. Problems such as the following were frequently encountered in the public examinations: (a) *Find the fifth root of* 0.1234; (b) *Find the value of*

$$\sqrt{\frac{3456}{1239} + \frac{0.7615}{0.2871}}.$$

In some schools, the use of slide rules (which are logarithmic in nature—they may be considered as analog devices based on the laws of logarithms) might have been taught. A few lucky students would have acquired slide rules for themselves and gloried in the wonderful computational power that they bestowed upon the user. (For several decades in a row, it was a common sight to see students of science and engineering go about with slide rules sticking out of their pockets; indeed, the slide rule was an indispensable part of one's luggage if one were a student of science or engineering.)

But those days have gone by; common logarithms and slide rules have been rendered utterly obsolete by the invention of the electronic calculator. For pedagogic reasons, calculators have not yet been made part of the school curriculum at the 10th standard level; they are currently permitted to be used in examinations at the 12th standard level, but only by a few examination boards (notably the CISCE). The reasons for this are understandable, for it is well known that computational ability tends to suffer once one gets to depend on a calculator. But the day must come, inevitably, when even this will change. At that point, common logarithms as a computational tool will cease to be a component of the school curriculum.

When this does take place, will logarithms as a topic in the school curriculum vanish? Not quite, for logarithms possess another significance which has nothing whatever to do with computation; advances in computational technology will have no implication on their inclusion in the curriculum. This book is essentially an elaboration of this statement. It is about the many ways in which logarithms arise in a natural and inevitable manner in science; in short, about the rich non-computational significance of logarithms.

The pace of the book is quite slow in the beginning; the first half of the book (Chapters 1–6) is about elementary topics—the laws of indices, the use of exponential notation in science, geometric series, the laws of logarithms, compound interest,...; computation too is discussed. In Chapters 7–9, we shall see how logarithmic scales of measurement arise quite naturally where human senses are involved; e.g., in the classification of sounds by their intensities, in the classification of stars by brightness, in the classification of earthquakes, and so on. Chapters 10–13, following this, lift the level of the book substantially and introduce some ideas which go beyond the school curriculum. In Chapter 10, the number e is studied, and in Chapters 11–12 the properties of the exponential and logarithmic functions are discussed—properties that make these functions so usable in science. In Chapter 13, an alternative treatment of the logarithmic function is presented. These chapters may be somewhat challenging for the young reader, but they must be read for a better appreciation of the logarithmic function.

Shailesh A. Shirali
2002

Acknowledgements

I wish to thank Universities Press for all the support that I have received from them in this publication venture.

Chapter 1

Multiplication

1.1 Early methods

We shall begin by describing the method known to some as "Egyptian multiplication" and to others as "Russian multiplication". Evidently, there are many claimants for this method! We shall illustrate the method with the computation of 11×17. Two rows are prepared as shown below; successive halving takes place in one row (giving the numbers 11, 5, 2, 1; observe that remainders are not retained) and successive doubling in the other (giving the numbers 17, 34, 68, 136).

HALVING	11	5	2	1
DOUBLING	17	34	68	136

Scanning through the HALVING row, we identify the entries which are *odd*, and compute the sum of the corresponding entries in the DOUBLING row. Here, we compute the sum $17 + 34 + 136$, as these are the numbers that correspond to 11, 5 and 1 in the HALVING row. The sum is 187, and this is the product we seek $(11 \times 17 = 187)$.

Similarly, to compute 47×91, we obtain a sequence of halves (47, 23, 11, 5, 2, 1; as earlier, remainders are not retained), and a sequence of doubles (91, 182, 364, 728, 1456, 2912). The only even number in the first sequence is 2, so we compute the sum $91 + 182 + 364 + 728 + 2912 = 4277$. Thus, $47 \times 91 = 4277$.

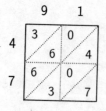

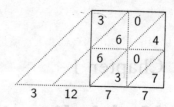

Figure 1.1. *Napier's grid*

EXERCISES

1.1.1 Compute 23×43 and 231×311 using this method.

1.1.2 Consider the computation of 16×103; let 16 be the number which is halved. What is the HALVING sequence in this case?

1.1.3 Consider the computation of 64×103; let 64 be the number which is halved. What happens in this case?

1.1.4 In the multiplication of 2-digit numbers by this technique, we find that we never need to add more than 7 numbers together. Why is this so?

1.1.5 Explain why the method works for the case when the number which we halve is of the form $2 \times 2 \times 2 \times \cdots \times 2$ (that is, a product of twos).

1.1.6 Explain why the method works in general.

1.2 Napier's bones

Another method is the 'grid' technique which probably originated in India early in the tenth century. We illustrate the technique with the computation of 47×91. The products of the individual digits are $4 \times 9 = 36$, $7 \times 9 = 63$, $4 \times 1 = 4$ and $7 \times 1 = 7$. The products go into an array as shown in Figure 1.1.

Reading down along the upward-sloping diagonals, left to right, the sums are 3, 12, 7 and 7. Working from right to left, we see that the required product is 4277 (the '1' in 12 has been carried and added to 3, giving 4). The transition from this technique to the one we currently use in schools to do long multiplication should be quite clear.

Centuries later, John Napier of Scotland patented a set of wooden rods that helped in these calculations. The device came to be known later as "Napier's bones".

EXERCISES

1.2.1 Compute 1235×571 using the grid method.

1.2.2 Make a set of 'Napier's bones' for yourself and use the set to do long multiplication.

1.3 Multiplication tables

Everyone is familiar with multiplication tables: $9 \times 1 = 9$, $9 \times 2 = 18$, $9 \times 3 = 27, \ldots$. However, tables of this sort have a serious limitation to them—they are just not big enough. What if one wanted to compute, say, 1234×5678? Would this require having to look up a table of products of 1234? And what about 14259×71326? It would be impossibly tedious and painful to have to build up so extensive a multiplication table as to be able to handle every conceivable multiplication—there would not be enough space to write down such a table! The early bankers, astronomers and navigators, who had to do vast numbers of hand-computations on a routine basis, certainly had an unenviable task on their hands.

Mathematicians realized a long time ago that the big difficulty with multiplication tables is that they have to be 2-dimensional; after all, multiplication needs *two* inputs. Mathematicians refer to such operations as *binary operations* ('binary' for 'two'). Suppose we needed a table for multiplication of numbers till 9. We would have to produce something like the following.

×	1	2	3	4	5	6	7	8	9
1	1	2	3	4	5	6	7	8	9
2	2	4	6	8	10	12	14	16	18
3	3	6	9	12	15	18	21	24	27
4	4	8	12	16	20	24	28	32	36
5	5	10	15	20	25	30	35	40	45
6	6	12	18	24	30	36	42	48	54
7	7	14	21	28	35	42	49	56	63
8	8	16	24	32	40	48	56	64	72
9	9	18	27	36	45	54	63	72	81

The reader will notice at once how large the array is; and this is for numbers till 9 only! The difficulty is, of course, that the array is 2-dimensional; it has to have $9 \times 9 = 81$ entries in all. Is it possible to make do with a 1-dimensional array instead?

Quite early on, it had been found that this *is* possible: simply by preparing a table of squares. The key observation is that the product $a \times b$ can be written as a difference of squares:

$$a \times b = \left(\frac{a+b}{2} \right)^2 - \left(\frac{a-b}{2} \right)^2.$$

Thus, to compute 6×14, we do the following:

$$\frac{14+6}{2} = 10, \quad \frac{14-6}{2} = 4,$$

$$10^2 - 4^2 = 100 - 16 = 84.$$

So the answer is: $6 \times 14 = 84$. This works for larger numbers as well, but we need to have a table of squares to refer to. For instance, to compute 1357×9131, we first compute

$$\frac{9131 + 1357}{2} = 5244, \quad \frac{9131 - 1357}{2} = 3887,$$

then we refer to the table of squares and look up

$$5244^2 = 27499536, \quad 3887^2 = 15108769,$$

and finally we do a subtraction: $27499536 - 15108769 = 12390767$. Therefore the answer is,

$$1357 \times 9131 = 12390767.$$

The reader should appreciate the saving achieved in having to prepare a 1-dimensional array as compared to a 2-dimensional array. (Note that we need to include the squares of half-integral numbers in our list, in order to multiply odd numbers with even numbers.

EXAMPLE. To do 61×72, we look up $66.5^2 = 4422.25$ and $5.5^2 = 30.25$, and obtain the answer as $4422.25 - 30.25 = 4392$.)

This takes care of multiplication, but what do we do about division, and what about the computation of powers and roots? Clearly, the method is limited in its utility. We need something that is a lot more flexible and powerful.

EXERCISES

1.3.1 Compute 211×319 and 171×352 using this technique. (You will need to refer to a table of squares.)

1.3.2 For multiplication of 3-digit numbers with one another, how large must the table of squares be? If we had to prepare an array that gave all possible products of 3-digit numbers with one another, how large would the array have to be? What saving is achieved by going in for the table-of-squares technique?

1.4 Interlude on prime factorization

The tabulation of a table of squares has an unexpected bonus—it helps in finding the prime factorization of numbers. The idea is very simple: if an integer N can be expressed as a difference of squares, say $N = a^2 - b^2$ where $a - b$ exceeds 1, then we have $N = (a-b)(a+b)$, and we have found the sought-after factorization. We show below how the idea can be put into practice.

Let the number to be factorized be $N = 1073$. We compute the quantities $N + 1^2$, $N + 2^2$, $N + 3^2$, ..., and continue till a square number is obtained. Obviously, we shall need to refer to a table of squares at each stage. Here is what we obtain.

$$N + 1^2 = 1074, \quad N + 2^2 = 1077,$$
$$N + 3^2 = 1082, \quad N + 4^2 = 1089 = 33^2.$$

We observe that $N + 4^2 = 33^2$. It follows that

$$N = 33^2 - 4^2 = (33 - 4)(33 + 4) = 29 \times 37,$$

and we have succeeded in factorizing N.

For $N = 2201$, we find that the first square in the sequence is

$$N + 20^2 = 2201 + 400 = 2601 = 51^2,$$

and this yields the factorization $N = 51^2 - 20^2 = 31 \times 71$.

EXERCISES

1.4.1 Factorize the numbers 2117 and 2911 using this technique.

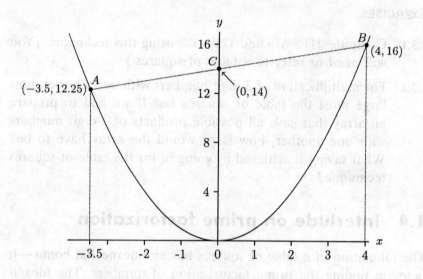

Figure 1.2. *Multiplication via the parabola $y = x^2$*

1.4.2 Find the factorization of 64777.

1.4.3 How good is the method? More precisely, which numbers require a small number of computations, and which numbers require a large number of computations?

1.5 Parabolic multiplication

We describe here a pretty, though perhaps not too practical, method for multiplying two numbers.

The method involves drawing the graph of a curve known as a *parabola*. We prepare a table of squares, then plot points with coordinates of the form (x, x^2) for numerous values of x; i.e., points with coordinates such as $(-4, 16)$, $(-3, 9)$, ..., $(0, 0)$, $(1, 1)$, ..., $(3, 9)$, $(4, 16)$, ... The resulting points fall neatly on a curve as shown in Figure 1.2. (Note that the scales are not the same: the scale on the y-axis is more compressed than the scale on the x-axis.)

Suppose we wish to multiply two numbers, say 3.5 and 4 (not much of a multiplication!—but it will do for the purposes of illustration). Points A and B are located on the curve as shown above: the x-coordinates of A and B are -3.5 and 4, respectively (note the negative sign for the x-coordinate of A), and segment

AB is drawn. Let AB meet the y-axis at C; then the y-coordinate of C yields the required product. From the sketch we see that the point C is $(0, 14)$, so the required product is 14.

More generally, to compute $a \times b$, we draw the segment connecting the points $A(-a, a^2)$ and $B(b, b^2)$; then the y-coordinate of the point C where AB meets the y-axis yields the desired product. The reason is easy to see. The slope of segment AB is

$$\frac{b^2 - a^2}{b - (-a)} = \frac{b^2 - a^2}{b + a} = b - a,$$

so if $C = (0, c)$, then by equating the slopes of BC and AB, we get

$$\frac{b^2 - c}{b - 0} = b - a, \quad \therefore \quad b^2 - c = b(b - a), \quad \therefore \quad c = ab.$$

So the y-coordinate of C is ab; hence the result.

1.6 Interlude on nomograms

The graphical approach described above may remind us of *nomograms*. These are graphical devices used to compute the values of a two-variable function f (which by definition needs two inputs; e.g., f could be addition or multiplication). In a nomogram you have three parallel scales, each with numbers written on it in some appropriate manner. For convenience, we refer to the top and bottom scales as the A and B scales, and to the middle one as the C scale. The nature of f decides the way the scales are to be graded. The manner of usage is this: to compute $f(a, b)$, we locate the numbers a and b on the A and B scales; hold a ruler so that its edge passes through the points a and b, then read off the answer from the point where the ruler meets the C scale. Nomograms are of great use on the engineering shopfloor, where particular kinds of computations need to be done repetitively. In cases where the computations are hard to do by hand, nomograms prove to be very convenient. We shall consider a few examples to render the idea more concrete.

- *Arithmetic mean.* Let f be the AM function given by $f(a, b) = (a + b)/2$. The three scales in this case are graded uniformly, all in the same way.

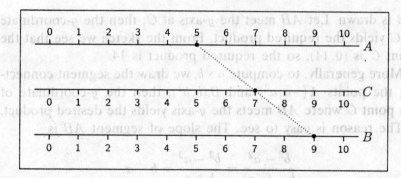

Figure 1.3. *A nomogram to compute the arithmetic mean*

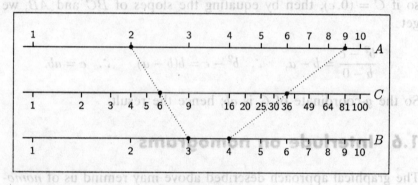

Figure 1.4. *A nomogram to compute products*

A 'ruler' has been drawn (Figure 1.3) to show how the AM of 5 and 9 is computed to be 7.

- *Product*. Let f be the product function given by $f(a, b) = ab$. The appropriate nomogram is shown in Figure 1.4. Observe that none of the scales have uniform gradation.

  A 'ruler' has been drawn to show how 2×3 is found to be 6, and how 9×4 is found to be 36.

Nomograms can be constructed to handle many kinds of two-variable functions, and their construction sometimes involves a lot of challenging and pretty mathematics.

★ ★ ★

Perhaps the most well-known form of the nomogram – though it is not generally referred to as such – is the slide rule, which was invented during the nineteenth century and, for many generations, was considered to be an indispensable tool for every working engineer and scientist. A popular stereotype of the first half (and more) of the twentieth century is the engineer with a slide rule sticking out of his pocket. Interestingly, when the famous geneticists Jim Watson and Francis Crick posed for their Nobel prize photograph in front of their bits-and-pieces double helical model of the DNA, Crick chose to do so with a slide rule in his hand, pointing it at the model in a rather theatrical manner.

But history has its own ironies; today the slide rule is no more, and few children (and even adults) have ever seen or handled one. Such is the power of the electronic revolution!

Chapter 2

Indices

2.1 The laws of indices

We shall begin by introducing the index notation (also known as the *exponential notation*). Consider the number $2 \times 2 \times 2$. Since 2 is being multiplied by itself 3 times, we write the number as 2^3, and we read the symbol aloud as "2 to the power 3" (or simply "2 to the 3"). The '3' which occurs above the '2' is called the *exponent*. So we have $2^3 = 8$, since 2 multiplied by itself 3 times yields 8. Similarly, we have the following relations:

$$2^4 = 2 \times 2 \times 2 \times 2 = 16,$$
$$2^5 = 2 \times 2 \times 2 \times 2 \times 2 = 32,$$
$$3^6 = 3 \times 3 \times 3 \times 3 \times 3 \times 3 = 729.$$

Why would we want to introduce such a notation? It is simply a matter of convenience—surely it saves time if we write 10^9 instead of 1000000000. It is one of the curious things about good notation that once one grows familiar with it, it starts to assume a life of its own and enables us to do things which we could not have done otherwise; it even helps in finding new results. This is something that has happened time and again in mathematics and science.

EXERCISES

2.1.1 List the powers of 2 from 2^1 till 2^{10}.

2.1.2 List the powers of 3 from 3^1 till 3^8.

2.1.3 Compute the quantities 4^5 and 7^3.

2.1.4 In each case, compare the two quantities listed below, and describe what you find.

(a) $10^2 \times 10^3$ and 10^5.

(b) $2^4 \times 2^5$ and 2^9.

(c) $3^7 \div 3^4$ and 3^3.

(d) $10^7 \div 10^3$ and 10^4.

(e) $7^5 \div 7^2$ and 7^3.

2.1.5 Which is greater, 2^8 or 3^5?

2.1.6 Which is greater, 2^{20} or 3^{12}?

2.1.7 Which is greater, 2^{11} or 3^7?

★ ★ ★

Having introduced a symbol, we must now learn how to use it. Consider the following relations:

$$2^3 \times 2^4 = (2 \times 2 \times 2) \times (2 \times 2 \times 2 \times 2) = 2^7,$$
$$3^2 \times 3^3 = (3 \times 3) \times (3 \times 3 \times 3) = 3^5.$$

It is not hard to see that when we do multiplication, exponents get *added*. We write the rule thus: *For any quantity a and any two whole number exponents m, n,*

$$a^m \times a^n = a^{m+n}.$$

Likewise, in division, exponents get *subtracted*. That is, if $m > n$, then

$$a^m \div a^n = a^{m-n}.$$

For instance, we have

$$\frac{a^4}{a^2} = \frac{a \times a \times a \times a}{a \times a} = \frac{a \times a}{1} = a^2,$$

$$\frac{a^5}{a^2} = \frac{a \times a \times a \times a \times a}{a \times a} = \frac{a \times a \times a}{1} = a^3,$$

and so on. If n, the index in the denominator, is greater than m, the index in the numerator, then we are left with a term in the denominator rather than the numerator; for example,

$$\frac{a^2}{a^4} = \frac{a \times a}{a \times a \times a \times a} = \frac{1}{a \times a} = \frac{1}{a^2}.$$

Next, consider the quantities $\left(2^4\right)^2$ and $\left(3^3\right)^2$. Since $2^4 = 16$ and $16^2 = 256 = 2^8$, while $3^3 = 27$ and $27^2 = 729 = 3^6$, we see that

$$\left(2^4\right)^2 = 2^8, \quad \left(3^3\right)^2 = 3^6.$$

We thus arrive at another rule: *For any quantity a and any whole number exponents m, n,*

$$\left(a^m\right)^n = a^{mn}.$$

EXERCISES

2.1.8 Compute the quantities $\left(2^5\right)^2$ and $\left(2^6\right)^2$.

2.1.9 Compute the quantities $\left(3^4\right)^2$ and $\left(3^3\right)^4$.

2.1.10 Place the numbers 2^{30}, 8^9 and 16^8 in ascending order.

2.1.11 Compare the two quantities $\left(a^m\right)^n$ and $\left(a^n\right)^m$.

2.2 Big numbers

Playing around with the index notation, one quickly begins to discover very big numbers. If we list the powers of 2, we find that 2^{20} is somewhat more than a million:

$$2^{20} = 1048576,$$

while 2^{30} has crossed the 'billion frontier':

$$2^{30} = 1073741824.$$

Further calculations get tedious, but with a computer to aid us, we make progress and find that

$$2^{100} = 1267650600228229401496703205376,$$

a 31-digit number! Still more impressive is the number 2^{200}; it turns out to have 61 digits!

Getting a feel of how large these numbers really are can be difficult, as they lie so completely outside our everyday experience. Here is an example to illustrate what we mean: *If we stack 10^8 coins into a single column, how tall is the column?* Taking a coin to be roughly 0.1 cm in width, we see that the column will be

$10^8 \times 0.1 = 10^7$ cm high. How high is this? Since 1 meter is 100 cm, while 1 km is 1000 meters, we see that 1 km is 100000 or 10^5 cm. This means that the column is 10^2 or 100 km in height!

★ ★ ★

Here is another example; this one is associated with a well-known story. The Shah of Persia was once challenged by a visitor to a game of chess (*shatranj*, as chess was known in earlier times). The visitor won, and the Shah asked what he desired as a reward. The visitor politely stipulated that he must receive 1 grain of rice for the corner square of the chess board, $2^1 = 2$ grains of rice for the square next to it, $2^2 = 4$ grains for the next square, $2^3 = 8$ grains for the next square, and so on. The Shah was amused at how little the visitor seemed to desire and agreed readily enough.

Let us compute, for fun, how much rice the visitor is actually owed. Since the board has 64 squares, he must receive 2^{63} grains of rice for the last square of the board. Let us check how large this number is. Bringing the computer to our aid, we find that

$$2^{63} = 9223372036854775808,$$

a 19-digit number. We see that 2^{63} is "almost" equal to 10^{19}. (This is written in mathematics as: $2^{63} \approx 10^{19}$.) Just how large a number is 10^{19}? How much rice does 10^{19} grains represent? Experimentation suggests that roughly 100 grains of rice will fill a spoon, and that a spoon will hold roughly 1 ml (of water or rice). Let the rice be kept in rooms measuring (in meters) $10 \times 10 \times 10$. The volume of such a room is 10^3 cubic meters or $10^3 \times 10^6 = 10^9$ ml (since 1 meter equals 10^2 cm). It follows that each room will hold $10^9 \times 10^2 = 10^{11}$ grains of rice. Since the visitor is owed 10^{19} grains, we need $10^{19} \div 10^{11} = 10^8$ such rooms. But this would require more rice than there is in the entire world!

EXERCISES

2.2.1 If I wish to train for long-distance running, would a morning jog of 10^4 strides per day be adequate?

2.2.2 A book has roughly 10^5 letters in it. Estimate the thickness of the book.

2.2.3 A sheet of paper 0.01 cm thick is folded fifty times in succession. How thick is the resulting wad of paper?

2.3 The four twos

Another 'big numbers' game is the one involving four twos: *Given four 2s, which is the largest number that one can make with them?* (We are permitted to concatenate them together and form numbers, as in 2222 or 222^2.) Here are a few such numbers:

$$2222, \quad 222^2, \quad 22^{22}, \quad 2^{222}, \quad \dots.$$

Which is the largest of these numbers? Clearly, $222^2 > 2222$. This can be shown, without making exact computations, as follows:

$$222^2 > 100^2 = 10000 > 2222.$$

Next, $22^{22} > 222^2$, for

$$22^{22} > 22^4 = \left(22^2\right)^2 = 484^2 > 222^2.$$

Which is larger: 2^{222} or 22^{22}? As earlier, one can figure this out without having to make exact computations:

$$22^{22} < 32^{22} = \left(2^5\right)^{22} = 2^{110} < 2^{222},$$

so 2^{222} is the larger number. Is this the best we can do? Not at all—we have another candidate, 2^{22^2}. This is the same as 2^{484}, so it is substantially bigger than 2^{222}.

But still better is the number $2^{2^{22}}$. This number is truly gargantuan! To see why, observe that $2^{22} = 4194304$, so $2^{2^{22}} = 2^{4194304}$. This turns out to be a number with 1262612 digits; writing it out in full would require a book with several thousand pages!!

EXERCISES

2.3.1   Which number is the largest: 444^4, 44^{44} or 4^{444}?

2.3.2   Which number is larger, 4^{44^4} or $4^{4^{44}}$?

Estimating the sizes of big numbers   You may wonder how we estimated the size of $2^{2^{22}}$, a number so stupendously large that even the most powerful computer available today would think twice before starting to compute it in full! Here is how we went about the task.

Note that $2^{10} = 1024$ has 4 digits, while $2^{20} = 1048576$ has 7 digits. Let us prepare the table shown below, listing the number of digits in 2^n for various values of n.

n	10	15	20	25	30	100	200
# digits in 2^n	4	5	7	8	10	31	61

We notice a pattern soon enough: *The number of digits in 2^n is roughly 1 more than* $0.3 \times n$. For example, 31 is 1 more than 0.3×100, and 61 is 1 more than 0.3×200.

This means that we can expect 2^{1000} to have roughly 301 digits, and 2^{10000} to have close to 3001 digits. Indeed, we are close— exact computations show that 2^{1000} has 302 digits, while 2^{10000} has 3011 digits. Continuing, we expect $2^{4194304}$ to have close to $1 + 0.3 \times 4194304$ or roughly 1258292 digits. Using somewhat better estimates, we find that the number of digits is 1262612.

EXERCISES

2.3.3 Estimate the number of digits in 2^{222}.

2.3.4 Given the following numbers:

$$5^{10} = 9765625,$$
$$5^{20} = 95367431640625,$$
$$5^{40} = 9094947017729282379150390625$$

(so 5^{40} is a 28-digit number), estimate the number of digits in 5^{55}, 5^{555} and $5^{5^{55}}$.

2.3.5 *Rumour mongering.* Rumours spread at an amazing speed. Suppose that you get hold of a fat, juicy rumour and that in the next 5 minutes you go out and tell it to two other people. Within the next 5 minutes they, in turn, go out and tell it to two others. And so on . . . : each person who hears the story tells it to two others within the next 5 minutes. How many people know the secret within half-an-hour? And how many within the next 2 hours?

2.3.6 Comment on how reasonable the above problem is for describing the actual spread of a rumour.

2.4 Negative exponents

Earlier on, we had stated that for any quantity a and any two whole number exponents m, n with $m > n$, the following division rule applies:

$$a^m \div a^n = a^{m-n}.$$

However, if we apply the rule to the case when n exceeds m, we encounter something puzzling. For instance, the rule states that $10^2 \div 10^3 = 10^{-1}$. But what does 10^{-1} mean? How can we multiply a number negative 1 times with itself?

As always in such situations, we have to take a step backwards in order to resolve the mystery. The question essentially is: What do we *mean* by 10^{-1}? If we succeed in giving it a meaning so that the statement $10^2 \div 10^3 = 10^{-1}$ is both correct and meaningful, then we shall have achieved a victory of sorts. But this is not too hard to do. Since $10^2 \div 10^3 = 1/10$, we agree to give the following meaning to 10^{-1}:

$$10^{-1} = \frac{1}{10^1}.$$

More generally, we give the following meaning to a^{-m} when m is a whole number exponent:

$$a^{-m} = \frac{1}{a^m}.$$

We have one more such "agreement" to make before proceeding— we must assign a meaning to a^0. Applying the rule to the division $a^m \div a^m$, we obtain $a^{m-m} = a^0$. On the other hand, it is obvious that $a^m \div a^m = 1$. So we give the following meaning to a^0:

$$a^0 = 1.$$

The "victory" we have achieved by the two agreements is this: *The rules $a^m \times a^n = a^{m+n}$ and $a^m \div a^n = a^{m-n}$ now hold for all integer values of m and n.* For example, we have

$$a^2 \div a^4 = a^{-2} = \frac{1}{a^2},$$

$$a^3 \div a^3 = 1 = a^0,$$

$$a^{-2} \times a^3 = a^1 = a,$$

and so on. So we have brought a certain consistency to the rules governing indices.

$\star \star \star$

The value of this victory is not to be underestimated. As stated earlier, a notation that has consistency and simplicity has a way of acquiring a life of its own, and when this happens, it brings great power into the hands of the mathematician. A nice example of this is provided by the place value system, which originated in India (possibly in Mohenjo-Daro) and later found its way into Europe via traders and scientists from Arabia. Reflect for a few moments on the ease with which we currently do computations such as 123×456 and $16827 \div 71$, and on the difficulty in carrying out the same computations using Greek numerals, and you will realize the great power of the system. (Try doing the 'sum' XVIII times MLXXIV and you will see what we mean!)

Algebra itself provides a particularly good example of the power of notation, for it enables one to make generalizations that would otherwise be impossible. Here is a nice example of such a generalization.

> **Theorem** *Let n be a positive integer. Then $x + y$ is a factor of $x^n + y^n$ when n is odd, but not when n is even.*

EXAMPLE. For $n = 3$ we have $x^3 + y^3 = (x + y)(x^2 - xy + y^2)$. The reader should check that $x + y$ is not a factor of $x^4 + y^4$.

The use of algebra is indispensable not only in proving such a statement, but also in formulating it in the first place. Here is another nice example.

> **Little Fermat Theorem** *Let p be any prime number and k any whole number, then $k^p - k$ is a multiple of p.*

EXAMPLE. Let $p = 7$ and $k = 3$; then $k^p - k = 3^7 - 3 = 2184 = 7 \times 312$. As earlier, the use of algebra is indispensable in stating and proving the theorem.

★ ★ ★

In the case of indices, the rules worked out above ultimately lead the way to logarithms. (Historically, things did not quite happen in this way, but that is a different story altogether. In Chapter 4, you will find an account of how John Napier hit upon the idea of logarithms, and how it gradually evolved into the modern idea.)

EXERCISES

2.4.1 Compute the quantities $2^3 \times 2^{-5} \times 2^4$ and $\left(10^{-2}\right)^3$.

2.4.2 Compute the quantity $\left(2^{-4}\right)^{-5} \div \left(4^{-2}\right)^{-4}$.

2.5 Fractional exponents

Having worked out the rules for whole-number exponents, can we
do the same for *fractional exponents* too? As earlier, the question
of meaning must first be resolved: we must be clear what we *mean*
by symbols such as $a^{1/2}, a^{1/3}, a^{3/5}$, and so on. However, it is not
too hard to come up with reasonable meanings for these symbols.

If we apply the rule $a^m \times a^n = a^{m+n}$ to $a^{1/2} \times a^{1/2}$, we obtain
a^1 (since $1/2 + 1/2 = 1$). So if we write b for $a^{1/2}$, then we have
$b^2 = a$. Surely, a quantity b for which $b^2 = a$ should by rights
be called the *square root* of a. (A square root of a is a quantity
which when squared ("raised to the power 2") yields a. We use
the symbol $\sqrt{a}$ to denote the square root of a.)

EXAMPLE. 2 is a square root of 4, and so is -2; 3 is a square root
of 9, and so is -3; 2.5 is a square root of 6.25; and so on.

Since we would like the rule $a^m \times a^n = a^{m+n}$ to apply to *all*
exponents, in particular to fractional exponents, we agree to regard
$a^{1/2}$ as the square root of a. Similarly, $a^{1/3}$ is the cube root of a,
for

$$a^{1/3} \times a^{1/3} \times a^{1/3} = a^{1/3+1/3+1/3} = a^1 = a,$$

and $a^{3/5}$ is the cube of the fifth root of a, for

$$\left(a^{3/5}\right)^5 = a^{(3/5)\cdot 5} = a^3.$$

More generally, $a^{1/n}$ is an "n^{th} root" of a, for we have the relation
$\left(a^{1/n}\right)^n = a^1 = a$, and $a^{m/n}$ is the mth power of the nth root of a.

EXAMPLE. $25^{1/2} = 5$, $100^{3/2} = 10^3 = 1000$, $512^{2/3} = 8^2 = 64$, and
so on.

★ ★ ★

The ambiguity between "a square root' and "the square root"
needs to be resolved before proceeding. We say "a square root",
because a number can have more than one square root; thus,
both 5 and -5 are square roots of 25. Generally, however, our
interest lies only in the positive square root, and when this is the
case, we refer to this root as "the square root". So we write "the
square root of 25 is 5", and so on. As we advance into higher

algebra, we find that a similar statement holds for the case of higher roots—cube roots, fourth roots, fifth roots, and so on; we talk of "the cube root" of a number, and also "a cube root" of a number. This happens when we study the so-called "complex numbers"; but more of that later.

★ ★ ★

The square root of 2 is approximately 1.4142; this is written as $\sqrt{2} \approx 1.4142$. Unfortunately, the exact value of $\sqrt{2}$ cannot be written down as a terminating decimal or a recurring decimal, because $\sqrt{2}$ cannot be expressed in the form a/b where a, b are whole numbers: $\sqrt{2}$ *is an irrational number.* Some examples of terminating decimals are 1.1, 3.23 and 5.237; the fractions corresponding to these numbers are $11/10$, $323/100$ and $5237/1000$, respectively. Some examples of recurring decimals are $0.1111\ldots$ (with infinitely many 1s); $0.12\,12\,12\ldots$ (with infinitely many repetitions of '12'); and $0.123\,123\,123\ldots$ (with infinitely many repetitions of '123'). The fractions corresponding to these decimals are $1/9$, $4/33$ and $41/333$ respectively.

In some cases the decimal does repeat, but only after an initial non-repeating portion.

EXAMPLE. The decimal $0.12\,345\,345\,345\ldots$ is such a number, and it corresponds to the fraction $12333/99900$.

All terminating and recurring decimals arise from fractions; they are also called *rational numbers.* As mentioned above, there are numbers such as $\sqrt{2}$ which are not of this type; their decimal expansions neither terminate nor eventually recur. Other examples of such numbers are $\sqrt{3}$ and $\sqrt[3]{2}$, and infinitely many such examples can be given.

EXERCISES

2.5.1 Compute the quantities $16^{5/2}$ and $36^{3/2}$.

2.5.2 Compute the quantities $27^{2/3}$, $125^{5/3}$ and $32^{6/5}$.

2.5.3 Compute the quantity

$$7^{1/4} \times 7^{3/4} \times 7^{5/4} \times 7^{7/4}.$$

2.5.4 Which is greater: $2^{1/2}$ or $3^{1/3}$?

2.5.5 Which is greater: $3^{1/4}$ or $5^{1/6}$?

2.5.6 Find the square roots of the following numbers: (a) 10^6 (b) 4×10^4 (c) 9×10^6 (d) 1.6×10^5 (e) 19.6×10^3.

2.5.7 Compute the values of

(a) $\left(3^2 + 4^2\right)^{1/2}$,

(b) $\left(3^3 + 4^3 + 5^3\right)^{1/3}$,

(c) $\left(4^3 + 4^2 + 4^0\right)^{1/2}$.

2.5.8 Find the values of

(a) $\sqrt{20\frac{1}{4}}$,

(b) $\sqrt{30\frac{1}{4}}$,

(c) $\sqrt{40\frac{1}{9}}$.

2.5.9 Find a positive integer a such that $a/2$ is a square and $a/3$ is a cube.

2.5.10 Find a positive integer a such that $a/5$ is a square, $a/2$ is a cube and $a/3$ is a fifth power.

2.5.11 Consider the following table, which lists the values of 2^x for $x = 0, 1, 2, \ldots$

x	0	0.5	1	1.5	2	2.5	3	3.5	4	...
2^x	1	1.414	2		4		8		16	...

Imagine that the table is filled in very finely with the values of 2^x listed for values of x such as $0.1, 0.2, \ldots$.

(a) Fill in the blanks in the table—that is, find 2^x for $x = 1.5, 2.5, 3.5, \ldots$.

(b) Suppose we wish to find x such that $2^x = 10$. Clearly, x must lie between 3 and 4 (because $2^3 = 8$ and $2^4 = 16$, and 10 lies between 8 and 16). Indeed, we can expect x to lie closer to 3 than to 4. How would you find the value of x?

(c) Find the approximate value of x if (a) $2^x = 5$ (b) $2^x = 9$.

2.5.12 Given that $\sqrt{3} \approx 1.732$, prepare a table listing the values of 3^x for $x = 0, 0.5, 1, 1.5, 2.5, \ldots$.

2.5.13 Find the approximate value of x if (a) $3^x = 10$ (b) $3^x = 20$ (c) $3^x = 50$.

2.6 Exponential notation in science

As emphasized several times earlier, a good notation has a way
of assuming a life of its own; at times, it helps users to think
of possibilities which otherwise they might never have considered
seriously. This being so, a simple test can be devised for measuring
how good a new system of scientific notation is: simply observe
how quickly it gets accepted by the scientific community, and how
it gets to be used in standard scientific writing. The exponential
notation passes the test with flying colours, and to verify this all
that one has to do is to glance through a science encyclopaedia.
The examples compiled below, giving data on miscellaneous topics,
testify eloquently to this fact.

Data on the Solar System   Given in Table 2.1 are various data
relating to each planet: r, its distance from the sun in km; T,
the time taken to complete a revolution around the sun (this is
its *orbital period*), measured in earth years; and D, its equatorial
diameter. Data on the Sun are given at the end.

Observe how easy it is to compare the data when it is given in
this form. For instance, comparing the Earth and Pluto, we see
that the Pluto–Sun distance is roughly 40 times the Earth–Sun
distance (for $587.2 \div 14.86$ is roughly $600 \div 15$ or 40). *The calculation
can be done mentally.*

Miscellaneous Physical Data   Listed below are many physical
constants used routinely by scientists. The commonly accepted
symbols are also given.

Table 2.1 *Miscellaneous Data on the Solar System*

Planet	r (km)	T (yrs)	D (km)
Mercury	5.75×10^6	0.24	3.46×10^3
Venus	10.75×10^6	0.62	4.64×10^3
Earth	14.86×10^7	1.00	1.22×10^4
Mars	22.65×10^7	1.88	6.72×10^3
Jupiter	77.32×10^7	11.86	1.39×10^5
Saturn	141.79×10^7	29.46	1.14×10^5
Uranus	285.28×10^7	84.01	4.70×10^4
Neptune	447.04×10^7	164.79	4.48×10^4
Pluto	587.20×10^7	248.43	?

- The speed of light in a vacuum: $c = 299792458 \approx 3 \times 10^8$ meter/second.
- The charge of an electron: $e = 1.6022 \times 10^{-19} \approx 1.6 \times 10^{-19}$ coulomb.
- Bohr radius: 5.2918×10^{-11} meter (this is a measure of the radius of the Hydrogen atom).
- Mass of an electron: $m_e = 9.1094 \times 10^{-28} \approx 9.1 \times 10^{-28}$ gram.
- Mass of a proton: $m_p = 1.6726 \times 10^{-24} \approx 1.67 \times 10^{-24}$ gram.
- Mass of a neutron: $m_n = 1.6749 \times 10^{-24} \approx 1.67 \times 10^{-24}$ gram.
- Classical radius of an electron: $r_e = 2.8179 \times 10^{-15} \approx 2.8 \times 10^{-15}$ meters.
- Avogadro constant: $A = 6.0221 \times 10^{23} \approx 6 \times 10^{23}$ per mole. (This gives the number of molecules of a gas per mole.)
- Mass of the Sun: 1.99×10^{30} kg.
- Radius of the Sun: 6.96×10^8 meters.
- Mass of the Earth: 5.98×10^{24} kg.
- Mass of Jupiter: 1.899×10^{27} kg.
- Mass of the Earth's Moon: 7.35×10^{22} kg.
- Radius of the Earth's Moon: 1.74×10^6 meters.
- Mass of the Solar System: 1.993×10^{30} kg.

Observe that more than 99.8% of the mass of the Solar System is contained in the Sun!

EXERCISES

2.6.1   Refer to the 'Solar System Data' table and do the following computations. For each planet, compute its distance (r) from the sun in Astronomical Units (AU, one AU being the distance of the Earth from the Sun). Plot a graph of r against the orbital period (T). Does the graph suggest any relationship between r and T?

2.6.2   What is the ratio of the volume of Jupiter to that of the Earth? How many Earths would fit into Jupiter? (Recall that the volume of a sphere is proportional to the *cube* of its radius.)

2.6.3 What fraction of the mass of the Solar System belongs to the Sun, and what fraction to Jupiter?

2.6.4 The Sun is made up almost entirely of hydrogen; it also has some helium, and traces of a few higher elements. Roughly, how many hydrogen atoms are there in the Sun?
(A hydrogen atom has one proton and one electron.)

2.6.5 If the Earth were made entirely of hydrogen, roughly how many atoms would be contained in the Earth?

2.6.6 What is the ratio of the mass of an electron to that of a proton?

2.6.7 Approximately, how many seconds are there in a year? Express your answer in the form 10^n.

2.6.8 What is your age in seconds? (Express your answer in the form 10^n.)

2.6.9 Taking the average human pulse rate to be 72 beats/minute and the average lifespan to be 75 years, roughly how many times does the heart beat during an average lifespan?

2.6.10 The *age of the universe* has been estimated as being about 2×10^{10} or 20 billion years. How many seconds old is the universe?

2.6.11 A *light year* is the distance travelled by a ray of light in one year. How many kilometers are there in a light year?

2.6.12 How many seconds does light take to travel from the Sun to the Earth, and how many seconds to travel to Pluto? What is the Earth–Sun distance in light minutes?

Chapter 3

Geometric Series

In this chapter, we shall study some properties of the sequence of powers of some fixed number a:

$$1, \ a, \ a^2, \ a^3, \ a^4, \ \ldots$$

Such series are referred to as *geometric series* or *geometric progressions*. We shall find that such series possess many features of interest. We shall also study two such series, corresponding to the cases $a = 2$ and $a = 3$. We consider, in particular, the properties found when we compute the cumulative sums of the series.

3.1 The powers of 2

Consider the sequence of powers of 2: $2^0, 2^1, 2^2, 2^3, \ldots$, or

$$1, \ 2, \ 4, \ 8, \ 16, \ 32, \ 64, \ \ldots.$$

We compute the sums of consecutive members of the sequence: $1, 1 + 2, 1 + 2 + 4, \ldots$; here is what we get:

$$1, \ 3, \ 7, \ 15, \ 31, \ 63, \ 127, \ \ldots.$$

We quickly notice that each number is 1 less than a power of 2; indeed, 1 less than the *next* higher power of 2. Thus, we guess that for any whole number n,

$$1 + 2 + 4 + \cdots + 2^{n-1} = 2^n - 1.$$

Please check that the formula 'works' when $n = 8$ and 9.

Computing the sums of the reciprocals of the powers of 2 also produces interesting results. Here, we consider the numbers:

$$1, \frac{1}{2}, \frac{1}{4}, \frac{1}{8}, \frac{1}{16}, \frac{1}{32}, \ldots$$

Their cumulative sums, i.e., the sums 1, $1 + 1/2$, $1 + 1/2 + 1/4$, $1 + 1/2 + 1/4 + 1/8$, $\ldots$, are the following numbers:

$$1, \frac{3}{2}, \frac{7}{4}, \frac{15}{8}, \frac{31}{16}, \frac{63}{32}, \ldots$$

Observe that the sums get closer and closer to 2: $7/4$ is $1/4$ short of 2, $15/8$ is $1/8$ short of 2, $31/16$ is $1/16$ short of 2, $63/32$ is $1/32$ short of 2, and so on. Indeed, the shortfall seems to get halved each time! We guess, therefore, that

$$1 + \frac{1}{2} + \frac{1}{4} + \frac{1}{8} + \cdots + \frac{1}{2^n} = 2 - \frac{1}{2^n}.$$

The interesting thing that this formula suggests is that if we compute the sum of the *infinite* series:

$$1 + \frac{1}{2} + \frac{1}{4} + \frac{1}{8} + \frac{1}{16} + \frac{1}{32} + \cdots \text{ (to infinity)},$$

then 'in the limit' we should get a sum of 2. This is because the shortfall from 2 (namely, $1/2^n$) gets smaller and smaller as n increases, and 'in the limit' it vanishes.

By using positive and negative signs in alternation another nice pattern emerges. Here, we compute the sums $1/2$, $1/2 - 1/4$, $1/2 - 1/4 + 1/8$, $1/2 - 1/4 + 1/8 - 1/16$, and so on. The sums are listed below:

$$\frac{1}{2}, \frac{1}{4}, \frac{3}{8}, \frac{5}{16}, \frac{11}{32}, \ldots$$

The numbers may not at first sight appear to have a simple pattern, but if we multiply each fraction by 3 and subtract 1, something nice emerges:

$$\frac{1}{2}, \ -\frac{1}{4}, \frac{1}{8}, \ -\frac{1}{16}, \frac{1}{32}, \ldots$$

These are just the reciprocals of the powers of 2, alternately plus and minus. We guess, therefore, that

$$\frac{1}{2} - \frac{1}{4} + \frac{1}{8} - \frac{1}{16} + \cdots \pm \frac{1}{2^n} = \frac{1}{3}\left(1 \pm \frac{1}{2^n}\right),$$

with the same sign (+ or −) on both sides, in place of the '±'. This yields yet another interesting result: if we compute the sum of the infinite series

$$\frac{1}{2} - \frac{1}{4} + \frac{1}{8} - \frac{1}{16} + \cdots \text{ (to infinity)},$$

then 'in the limit' we get a sum of 1/3.

EXERCISES

3.1.1 Verify that $1 + 2 + 2^2 + 2^3 + \cdots + 2^{10} = 2^{11} - 1$.

3.1.2 Show from the relation

$$1 + 2 + 4 + 8 + \cdots + 2^{n-1} = 2^n - 1,$$

how one can deduce that

$$1 + \frac{1}{2} + \frac{1}{4} + \frac{1}{8} + \cdots + \frac{1}{2^{n-1}} = 2 - \frac{1}{2^n}.$$

3.1.3 Show that the sum

$$1 - \frac{1}{2} + \frac{1}{4} - \frac{1}{8} + \frac{1}{16} - \cdots \text{ (to infinity)}$$

equals 2/3.

3.2 The powers of 3

We do the same thing with the powers of 3 and quickly find that here too many features of interest are to be found. Here are the first few powers: $1, 3, 3^2, 3^3$, or

$$1, 3, 9, 27, 81, 243, \ldots$$

The cumulative sums are

$$1, 4, 13, 40, 121, 364, \ldots$$

These numbers may not at first sight seem to have any simple pattern, but if we double each number, we get the following sequence:

$$2, 8, 26, 80, 242, 728, \ldots$$

We quickly notice that each number here is 1 *less* than a power of 3. We guess, therefore, that

$$2\left(1 + 3 + 9 + 27 + \cdots + 3^{n-1}\right) = 3^n - 1.$$

As earlier, we may also work with the reciprocals of the powers of 3. Here is what we find: the sums 1, $1 + 1/3$, $1 + 1/3 + 1/9$, $1 + 1/3 + 1/9 + 1/27$ are

$$1, \ \frac{4}{3}, \ \frac{13}{9}, \ \frac{40}{27}, \ \frac{121}{81}, \ \ldots$$

Can you spot the pattern in this series of fractions? What sum is obtained 'in the limit' by adding infinitely many fractions of the series?

EXERCISES

3.2.1 Verify that $1 + 3 + 9 + 27 + \cdots + 3^9 = \left(3^{10} - 1\right)/2$.

3.2.2 Verify that the sum

$$1 + \frac{1}{3} + \frac{1}{9} + \frac{1}{27} + \frac{1}{81} + \cdots \text{ (to infinity)}$$

is equal ('in the limit') to $3/2$.

3.2.3 Verify that the sum

$$1 - \frac{1}{3} + \frac{1}{9} - \frac{1}{27} + \frac{1}{81} - \text{ (to infinity)} \cdots$$

is ('in the limit') equal to $3/4$ (or half the sum obtained when the signs are all $+$).

3.2.4 Find a formula which connects the sum

$$1 + 10 + 10^2 + 10^3 + \cdots + 10^{n-1}$$

with 10^n.

3.2.5 Verify that the sum

$$1 + \frac{1}{10} + \frac{1}{100} + \frac{1}{1000} + \cdots \text{ (to infinity)}$$

is equal ('in the limit') to $10/9$.

3.2.6 Verify that the sum

$$1 - \frac{1}{10} + \frac{1}{100} - \frac{1}{1000} + \frac{1}{10000} - \cdots \text{ (to infinity)}$$

is equal ('in the limit') to 10/11.

3.2.7 Let a be any given number. Guess a formula which connects the sum

$$1 + a + a^2 + a^3 + \cdots + a^{n-1}$$

with a^n.

3.2.8 Let a be a number greater than 1. Guess a formula (in terms of a) for the sum

$$1 + \frac{1}{a} + \frac{1}{a^2} + \frac{1}{a^3} + \frac{1}{a^4} + \cdots \text{ (to infinity)}.$$

3.3 Recurring decimals

The problems posed above (in the preceding section) immediately bring to mind the topic of recurring decimals. Consider the fraction 1/9. If we divide 1 by 9, we find that the division never terminates, and we obtain an infinite decimal

$$\frac{1}{9} = 0.1111111\ldots$$

If we divide 1 by 11, we obtain another such infinite decimal

$$\frac{1}{11} = 0.0909090909\ldots$$

Such decimals are called *recurring decimals*, for they consist of infinitely many repetitions of some block of digits. Many such instances can be constructed; for example,

$$\frac{1}{7} = 0.142857\,142857\,142857\ldots.$$

Here the repeating portion is the block '142857'.

NOTE. In some cases, the repetitive portion does not start immediately after the decimal point. For example, we have $1/6 = 0.1666\ldots$ (the digit '6' repeats indefinitely), $4/15 = 0.2666\ldots$, and so on.

EXERCISES

3.3.1 Find the repeating blocks for the decimals obtained from the fractions $1/3$, $1/13$ and $1/17$.

3.3.2 Show how the statements $0.11111\ldots = 1/9$ and

$$\frac{1}{10} + \frac{1}{100} + \frac{1}{1000} + \cdots \text{ (to infinity)} = \frac{1}{9}$$

are different ways of writing the same thing.

3.3.3 Show how $0.01010101\ldots = 1/99$ and

$$\frac{1}{100} + \frac{1}{10000} + \frac{1}{1000000} + \cdots \text{ (to infinity)} = \frac{1}{99}$$

are different ways of writing the same thing.

3.3.4 Let $a = 18$ and $b = 100 = 10^2$. Find the fraction that corresponds to the sum

$$\frac{a}{b} + \frac{a}{b^2} + \frac{a}{b^3} + \frac{a}{b^4} + \ldots \text{ (to infinity)}.$$

(Observe that the question asks for the fraction whose decimal representation is $0.18\,18\,18\,18\ldots$.)

3.3.5 Let $a = 142857$ and $b = 1000000 = 10^6$. Find the fraction that corresponds to the sum

$$\frac{a}{b} + \frac{a}{b^2} + \frac{a}{b^3} + \frac{a}{b^4} + \ldots \text{ (to infinity)}.$$

3.3.6 Find the fraction whose decimal representation is the infinite decimal $0.123\,123\,123\,123\ldots$. In other words, find the fraction which corresponds to the sum

$$\frac{a}{b} + \frac{a}{b^2} + \frac{a}{b^3} + \frac{a}{b^4} + \ldots \text{ (to infinity)},$$

when $a = 123$ and $b = 1000$.

3.4 Achilles and the tortoise

The ideas developed in the preceding section help in resolving a famous paradox that goes back to Greek times, about a race that Achilles had with a tortoise. Assume that Achilles runs at a speed of 10 meters per second (this makes him about as fast as Carl Lewis, the 100-meter Gold Medalist of the 1984 and 1988 Olympics) while the tortoise runs at 1 meter per second (so Achilles runs 10 times as fast as the tortoise). Let the tortoise be given a 100 meter headstart over Achilles. The question is: *Does Achilles ever catch up with the tortoise?*

"Of course he does!" is your immediate answer. But how do you counter the following piece of reasoning? At the start, the tortoise is 100 meters ahead of Achilles. During the time that Achilles takes to cover this distance, the tortoise moves forward by 10 meters. During the time that Achilles takes to cover this distance (the 10 meters), the tortoise moves forward by another 1 meter. During the time that Achilles takes to cover *this* distance, the tortoise moves forward by another 0.1 meter. So the tortoise always seems to be ahead of Achilles! Each time that Achilles covers the deficit, the tortoise has advanced a little further. So by this logic, Achilles will never catch up with the tortoise! Where has our reasoning gone wrong?

EXERCISES

3.4.1 The time taken by Achilles to cover the initial headstart of 100 meters is 10 seconds; the time needed to cover the next 10 meters is 1 second; the time needed to cover the next 1 meter is 0.1 second; and so on. Thus, the time in seconds that Achilles should take in catching up with the tortoise is

$$10 + 1 + 0.1 + 0.01 + 0.001 + \cdots \text{ (to infinity)}.$$

What is the sum of this infinite series?

3.4.2 Note the vitally important point that the limiting sum of the series displayed above is *not* infinite. This is precisely where the Greeks went wrong—namely, in assuming that the sum of infinitely many quantities must itself be infinite. So, after all, Achilles does catch up with the tortoise!

Look for other instances where reasoning of the type used by the Greeks leads to absurd conclusions. As a sample, we present one more such paradox.

Suppose that I have to move from a point A to a point B, a distance of 100 meters; will I ever get to B? Well ... I have to first reach C, the point half-way between A and B. Next, I have to reach D, the point half-way between C and B. Following this, I have to reach E, the point half-way between D and B; and so on. So, with infinitely many moments in-between at which I have not reached B, how can I ever hope to reach B?

(Of course, in this problem, I will also run out of alphabets!)

3.4.3 *A Population Paradox.* I have 2 parents; each parent has 2 parents, so I have $2^2 = 4$ grandparents. Each grandparent has 2 parents, so I have $2^3 = 8$ great-grandparents. Next, each great-grandparent has 2 parents, so I have $2^4 = 16$ great-great-grandparents. Continuing, I conclude that n generations ago I had 2^n ancestors. In particular, 50 generations ago, I had 2^{50} ancestors. This is a *very* large number; it exceeds one thousand trillion, which is many times the total population on the Earth today! How could I possibly have had so many ancestors, and how could the Earth have had more human beings in the past than it has today? What is wrong with my reasoning?

3.4.4 *The Bee and the Cyclists.* Let two cyclists, A and B, start from a point 100 km apart; each cycles at a speed of 10 km per hour. As they start, a bee takes off from the nose of cyclist A and heads towards B's nose, proceeding at 20 km per hour. The instant it alights on B's nose, it turns back and heads towards A's nose. It continues this back and forth movement, doing shorter and shorter stretches until finally it gets squashed between the noses of the two cyclists. How much distance does the bee travel before the Big Bang comes?

Chapter 4

Logarithms

4.1 Definition

Having found the laws governing indices, we soon discover some unexpected gains. For instance, consider the table below, which gives the powers of 2.

$2^0 = 1$	$2^4 = 16$	$2^8 = 256$	$2^{12} = 4096$
$2^1 = 2$	$2^5 = 32$	$2^9 = 512$	$2^{13} = 8192$
$2^2 = 4$	$2^6 = 64$	$2^{10} = 1024$	$2^{14} = 16384$
$2^3 = 8$	$2^7 = 128$	$2^{11} = 2048$	$2^{15} = 32768$

Having constructed the table, certain types of multiplications and divisions can be done quite rapidly, simply by looking up the appropriate entries in the table. For example,

$$64 \times 256 = 2^6 \times 2^8 = 2^{14} = 16384,$$
$$32768 \div 256 = 2^{15} \div 2^8 = 2^7 = 128,$$

and so on. Clearly, if we had a "complete list" of such powers, then we would be able to handle any multiplication or division problem.

The fact that we chose to use the powers of 2 has no particular significance; we could have done just the same with a table of powers of, say, 3 or 4 or 5. This thought more or less prompts the definition of the logarithm function. Let a be a fixed positive number, $a \neq 1$, and let b, c be numbers such that

$$a^c = b.$$

32

Then c is called the *logarithm of b to base a*, and we write

$$\log_a b = c.$$

That is, the logarithm of b to base a is the power to which a must to be raised in order to 'get' b. For instance, $\log_2 8 = 3$ since $2^3 = 8$, and $\log_2 16 = 4$ since $2^4 = 16$. Similarly, we have $\log_3 9 = 2$, $\log_3 27 = 3$, $\log_{10} 100 = 2$, and so on.

The logarithm need not be a whole number; for example, $\log_4 2 = 1/2$ because $4^{1/2} = 2$; $\log_9 3 = 1/2$ because $9^{1/2} = 3$; $\log_{16} 8 = 3/4$, because $16^{3/4} = 8$, and so on.

The vital thing to note is that once the base a is fixed ($a > 0$, $a \neq 1$), every number b has some definite logarithm. For example, let the base be 2. Consider the following sequence of numbers (the powers of 2) which extends infinitely far in both directions:

$$\ldots, \ 2^{-5}, \ 2^{-4}, \ 2^{-3}, \ 2^{-2}, \ 2^{-1}, \ 2^0, \ 2^1, \ 2^2, \ 2^3, \ 2^4, \ 2^5, \ \ldots,$$

or

$$\ldots, \frac{1}{32}, \frac{1}{16}, \frac{1}{8}, \frac{1}{4}, \frac{1}{2}, \ 1, \ 2, \ 4, \ 8, \ 16, \ 32, \ \ldots.$$

Observe how the numbers grow from being indefinitely small in size (nearly 0) to indefinitely large. Observe also that the powers are all positive. If the exponents are allowed to take values that are not whole numbers, then the resulting values span the entire range of numbers from 0 upwards (but not 0 itself). This means that every positive number has *some* logarithm to base 2. Clearly, the logarithm is negative if the number lies between 0 and 1 and positive if the number exceeds 1. (More generally, if the base a exceeds 1, then $\log_a x$ is negative for $0 < x < 1$ and positive for $x > 1$.)

Observe also that the logarithm of 1 to any positive base a is 0, because $a^0 = 1$ for every $a > 0$. (This was discussed in an earlier chapter.) Also observe that if $\log_a b = c$ (that is, $a^c = b$), then

- $\log_a b^2 = 2c$, because $a^{2c} = (a^c)^2 = b^2$;

- $\log_a b^3 = 3c$, because $a^{3c} = (a^c)^3 = b^3$;

- $\log_a b^n = nc$, because $a^{nc} = (a^c)^n = b^n$ for any $n \in \mathbf{N}$ (here $\mathbf{N}$ refers to the set of positive integers);

- $\log_a 1/b = -c$, because $a^{-c} = 1/a^c = 1/b$;

- $\log_a 1/b^2 = -2c$, because $a^{-2c} = (1/a^c)^2 = 1/b^2$;

and so on. To put it briefly,

$$\log_a b^k = k \log_a b,$$

and this holds for all k and all positive a, b. The other laws governing the use of logarithms follow naturally from the laws governing indices. Thus, we have

$$\log_a b + \log_a c = \log_a bc, \quad \log_a b - \log_a c = \log_a \frac{b}{c}.$$

To put it in words: *The operation of taking logarithms converts products into sums and quotients into differences.*

Note that we do not define logarithms for base 0 or 1, or for negative bases; nor do we define the logarithms of negative numbers.[1]

★ ★ ★

You may have noticed that in defining the idea of a logarithm, we essentially *inverted* the exponential relation. This kind of inversion is done quite often in mathematics. For instance, from the relation $a^2 = b$, we obtain, by inversion, the notion of the square root: $a = \sqrt{b}$. Likewise, from $a^3 = b$ we obtain the notion of the cube root: $a = \sqrt[3]{b}$; and from functions such as $y = \sin x$ and $y = \cos x$, we obtain, by inversion, the notion of the inverse trigonometric functions: $x = \arcsin y$ and $x = \arccos y$ (also written as $x = \sin^{-1} y$ and $x = \cos^{-1} y$, respectively). Naturally, when we invert the logarithmic relation, $y = \log x$, we obtain the exponential relation. For this reason, the exponential function is also known as the 'antilog' function. Thus, the following statements carry the same meaning:

$$\log_a b = c, \quad a^c = b, \quad \operatorname{antilog}_a c = b.$$

EXERCISES

4.1.1 Compute the values of $\log_4 16$ and $\log_5 125$.

4.1.2 Compute the values of $\log_9 3$ and $\log_{32} 16$.

4.1.3 Compute the sum $\log_5 1 + \log_5 3 + \log_5 5 + \log_5 7 + \log_5 9$.

---

[1]Actually this can be done; but to do so would take us into the realm of complex numbers, which is too advanced a topic to bring in at this point.

4.1.4 Show that $\log_a b = 1/\log_b a$.

4.1.5 Find a relation between $\log_a b$ and $\log_{1/a} 1/b$.

4.1.6 Show that

$$\frac{\log_a b}{\log_a c} = \log_c b.$$

4.1.7 Let $a = k^m$ and $b = k^n$, where a, b, k are positive numbers. Express $\log_a b$ in terms of m and n.

4.1.8 State the logarithmic equivalents of the relations: $A = \pi r^2$, $V = 4\pi r^3/3$.

4.1.9 Show that the nth root of a is equal to the antilogarithm of $(\log a)/n$. (The logarithm and antilogarithm are to the same base.)

4.1.10 Let a, b be positive numbers. Show that

$$a^{\log b} = b^{\log a},$$

the logarithms on both sides being to the same base.

4.1.11 Scientists sometimes use 'log plots' and 'loglog plots' when they draw graphs of their data. Look up the meanings of these terms. How does log plot graph paper and loglog plot graph paper differ from normal graph paper?

4.2 Genesis of the logarithm

The power and importance of the logarithm lie in the fact that it converts a product into a sum and a multiplication problem into an addition problem, and (similarly) a quotient into a difference and a division problem into a subtraction problem. This obviously has computational significance. We saw earlier, in Chapter 1, how a table of squares helps in changing a problem of multiplication into one of addition. This was known during John Napier's time; so were certain standard results from trigonometry that converted products into sums. It seems likely that it was such facts that helped Napier hit upon the idea of logarithms. The exact route that he took is unfortunately not known, but in the end it took the following form.

Consider a ray ℓ (see Figure 4.1), with A as the endpoint, and a line segment BC of unit length. Let particles X and Y start from

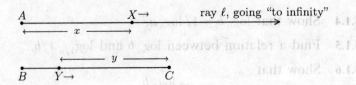

Figure 4.1. *Genesis of the Naperian logarithm*

A and B and move along ℓ and BC respectively, starting with the same initial speed; let X move at a constant speed, and let Y move at a speed proportional to the distance YC. This means that its speed decreases steadily as it approaches C, and it takes infinitely long to reach its destination.

Write x and y for the distances AX and YC, respectively. It is clear that x and y are related in a 1–1 manner: when x is 0, y is 1, and as x increases to infinity, y decreases to 0. The relationship is denoted by writing

$$x = \text{Nap} \log y.$$

Here 'Nap log' stands for 'Naperian logarithm'.

What is the relation of the 'Nap log' function to the one we know today as the logarithmic function? Let the length of BC be b units. Using the methods of the calculus, it can be shown that

$$\text{Nap} \log y = -\ln y = -\log_e y = \log_e 1/y,$$

where $e \approx 2.71828$ is the number known as Euler's number.

★ ★ ★

For those familiar with calculus, we shall provide justification for the above claim as follows. Observe that

$$\frac{dx}{dt} = 1, \quad \therefore \quad x = t \text{ (since } x = 0 \text{ when } t = 0\text{); and}$$

$$\frac{dy}{dt} = -y, \quad \therefore \quad t = -\ln y \text{ (since } y = 1 \text{ when } t = 0\text{).}$$

It follows that $x = -\ln y$.

★ ★ ★

Napier publicized his invention in 1614, in a book titled *A Description of the Wonderful Law of Logarithms*. This contained a table of Naperian logarithms (Napier took the length of BC to be

10^7 units rather than 1 unit to avoid having to deal with too many decimals). The significance of the new invention was quickly seen. In 1615, John Briggs of Cambridge visited Napier in Edinburgh (Scotland), and the two scholars agreed to modify the definition so that the logarithm of 1 would be 0 while the logarithm of 10 would be 1. Such logarithms are called 'common logarithms'. Shortly thereafter, Briggs published the first table of common logarithms (accurate to 14 decimal places, and computed by hand!), and this gained almost immediate acceptance in Europe. To astronomers in particular, who routinely had to perform the most difficult and complicated calculations, this device came as a boon.

★ ★ ★

In conclusion, here is a quote from F Cajori's *A History of Mathematics*.

> The miraculous powers of modern calculation are due to three inventions: the Arabic notation, Decimal Fractions and Logarithms. The invention of logarithms in the first quarter of the seventeenth century was admirably timed, for Kepler was then examining planetary orbits, and Galileo had just turned the telescope to the stars. During the Renaissance German mathematicians had constructed trigonometrical tables with great accuracy, but its greater precision enormously increased the work of the calculator. It is no exaggeration to say that the invention of logarithms "by shortening the labours doubled the life of the astronomer".
>
> Logarithms were invented by John Napier (1550–1617), Baron of Merchiston, in Scotland. It is one of the great curiosities of the history of science that Napier constructed logarithms before exponents were used. To be sure, Stifel and Stevin had made attempts to denote powers by indices, but this notation was not generally known—not even to T Harriot, whose ALGEBRA appeared long after Napier's death. That logarithms flow naturally from the exponential symbol was not observed until much later.

4.3 Brief historical details

The following historical details may be of interest. (See (Mao2) for a more extended account.) John Napier was born in 1550 or thereabouts in Merchiston, Scotland, and was a student of theology. He seems to have had a fervent interest in religious activity and wrote extensively: on anti-Catholic matters, on the Pope, on the Day of Judgement. On the whole, Napier comes across as a man with very diverse interests; also, unfortunately, as a rather quarrelsome man. (The two traits often go together! Let us be more optimistic and describe him simply as a rather colourful personality). He seems to have had a mechanical bent of mind—he invented a hydraulic screw for controlling water levels in coal pits, and he even tried his hand at designing artillery machines. But the verdict of history on Napier has been curious; if he is at all remembered, these many centuries later, it is because of another invention altogether—that of logarithms.

As stated above, the exact route taken by Napier in formulating the laws of logarithms is unknown to us. However, formulas such as

$$ab = \left(\frac{a+b}{2}\right)^2 - \left(\frac{a-b}{2}\right)^2,$$

and

$$\sin A \cdot \sin B = \frac{\cos(A-B) - \cos(A+B)}{2}$$

were well known in his time; both accomplish the useful purpose of converting products into sums, which are obviously simpler to compute. It is likely that it was just such rules that provided the impetus to Napier.

The other possibility is that he proceeded as we did earlier in the chapter—through the arithmetical theory of indices, which had been formulated earlier by Simon Stevin and Michael Stifel. (However, historians of mathematics are agreed that this did not happen; see Cajori's comment, above.) *If* at all it happened this way, here is a possible reconstruction: Napier realizes that if we construct a table of powers of some fixed number, then the table would provide a convenient scheme for multiplication.

EXAMPLE. Suppose we construct a table of powers of 2. Then, if we wish to multiply two positive numbers a and b, we consult the table to see where a and b are located and find, say, that $a = 2^c$ and $b = 2^d$. We compute the sum $c + d$, then consult the table once again and look up the entry corresponding to $c + d$ (the table is now used in the 'reverse' way).

Now the powers of 2 grow at a terrific pace, and there are large gaps between successive powers, so choosing 2 as a base is not too convenient. The use of fractional exponents will, of course, fill up the gaps, but in those times the use of decimal fractions was not too well known. Napier's choice is to use $1 - 10^{-7}$ or 0.9999999 as a base. This may seem a most peculiar choice, but it is done with a view to minimize the use of decimal fractions.

The base b having been decided, Napier must now compute a table of the powers of b. To avoid decimals he computes a table of values of $10^7 \times b^k$ or $10^7 \times (1 - 10^{-7})^k$, where k takes varying integer values. (The factor 10^7 keeps the decimals out.) This is an utterly tedious and exhausting task! Napier calls the exponent k the *logarithm* or *ratio number* of $10^7 \times b^k$. So Napier's definition of a logarithm amounts to this: *If $N = 10^7 \times (1 - 10^{-7})^k$, then k is the logarithm of N.*

★ ★ ★

This account may be fictitious; but whatever Napier's route, the task took him twenty years (or more) to complete. It was John Briggs (1561–1631) who proposed to Napier, when they met in Scotland, that the definition be modified so that the logarithm of 1 would be 0, while that of 10 would be 1. Napier agreed, and thus was born the common logarithm, whose base is 10.

The really remarkable thing about the genesis of logarithms is that the idea seems to have sprung up "from nowhere"; there is absolutely nothing in the mathematics of the time that might have suggested it.

4.4 Estimation of logarithms

This section could otherwise be titled CLOSE ENCOUNTERS OF THE POWERFUL KIND, for the basic idea in this section over and over again is to use "close encounters" between the power sequences of different integers. This may sound mysterious, but actually is readily explained.

Suppose that we wish to estimate the value of $\log_{10} 2$. We start by listing the powers of 2 and identifying one that is 'close' to a power of 10. The first such instance is $2^{10} = 1024$. Since 1024 is fairly close to 1000 or 10^3 (in the sense that the ratio of 1024 to 10^3 is close to 1), we write $2^{10} \approx 10^3$. Taking the 10th roots on both sides, we obtain $2 \approx 10^{0.3}$. This means that $\log_{10} 2 \approx 0.3$. This is a rather good approximation; a more accurate value, computed by using superior methods, is 0.3010299.

It turns out that a closer encounter occurs somewhat further down the sequence—we find that 2^{103} is close to 10^{31}:

$$2^{103} = 10141204801825835211973625643008 \approx 10^{31},$$

yielding $\log_{10} 2 \approx 31/103 = 0.30097\ldots$. Comparing this with the true value given above, we see that the error in the estimate is less than 0.02%.

Moving to the estimation of $\log_{10} 3$, we find the following close encounter:

$$3^{21} = 10460353203 \approx 10^{10}.$$

Other instances of closeness are 3^{109} (which starts with the digits $10144\ldots$ and is close to 10^{52}; it is a 53-digit number), and 3^{153} (which starts with the digits $998968\ldots$ and is close to 10^{73}; it is a 73-digit number). So we obtain the following estimates for $\log_{10} 3$:

$$\frac{10}{21} \approx 0.47619, \qquad \frac{52}{109} \approx 0.47706, \qquad \frac{73}{153} \approx 0.477124.$$

The closest encounter amongst the ones listed is the last one, so we expect the best estimate of these to be 0.477124. More advanced methods yield $\log_{10} 3 = 0.477121\ldots$, so our estimate differs from the true value only in the sixth decimal place.

Another instance of closeness is provided by the pair 7^{510} and 10^{431} (we find that 7^{510} is a 432-digit number that begins with the digits $100000093\ldots$). This yields:

$$\log_{10} 7 \approx \frac{431}{510} = 0.845098039215\ldots.$$

The true value is

$$\log_{10} 7 = 0.845098040014\ldots,$$

and we see that the two values differ only in the ninth decimal place. Considering how close the two numbers 7^{510} and 10^{431} are, the closeness of the estimate does not come as a surprise.

We do not have to restrict ourselves to logarithms to base 10. For instance, $5^2 = 25$ is rather close to $3^3 = 27$, which shows that $\log_3 5 \approx 3/2 = 1.5$. However, this is a poor approximation (the true value of $\log_3 5$ is $1.46497\ldots$).

With some ingenuity, one can obtain many such estimates. For instance, since $3^4 = 81$, which is close to $80 = 2^4 \times 5$, we see that $3^4 \approx 2^4 \times 5$, or

$$4 \log_{10} 3 \approx 4 \log_{10} 2 + \log_{10} 5.$$

Since $\log_{10} 5 = 1 - \log_{10} 2$, this yields:

$$4 \log_{10} 3 \approx 3 \log_{10} 2 + 1.$$

Substituting 0.3010 for $\log_{10} 2$, we obtain an estimate for $\log_{10} 3$:

$$\log_{10} 3 \approx \frac{1.9010}{4} = 0.47525.$$

Likewise, from the closeness of $11^2 = 121$ and $120 = 2^3 \times 3 \times 5$, we see that

$$
\begin{aligned}
2 \log_{10} 11 &\approx 3 \log_{10} 2 + \log_{10} 3 + \log_{10} 5 \\
&= 2 \log_{10} 2 + \log_{10} 3 + 1 \\
&\approx 0.6020 + 0.4771 + 1 = 2.0791,
\end{aligned}
$$

yielding $\log_{10} 11 \approx 1.03955$. A better estimate is 1.04139.

EXERCISES

4.4.1 Estimate the values of $\log_3 5$ and $\log_5 11$, using the 'close encounters' technique.

4.4.2 Which is larger: $\log_2 3$ or $\log_4 8$?

4.4.3 Which is larger: $\log_2 3$ or $\log_5 10$? (Hint: Search for a fraction that lies between the two quantities.)

4.4.4 Which is larger: $\log_3 7$ or $\log_5 16$?

4.5 Logarithms to base 2

Let us graphically show what logarithms to base 2 "look" like, and how they may be estimated. We start by constructing a table of the integral powers of 2.

$2^0 = 1$	$2^4 = 16$	$2^8 = 256$	$2^{12} = 4096$
$2^1 = 2$	$2^5 = 32$	$2^9 = 512$	$2^{13} = 8192$
$2^2 = 4$	$2^6 = 64$	$2^{10} = 1024$	$2^{14} = 16384$
$2^3 = 8$	$2^7 = 128$	$2^{11} = 2048$	$2^{15} = 32768$

These give us facts such as $\log_2 2 = 1$, $\log_2 4 = 2$, $\log_2 8 = 3$, and so on.

We now start to "fill in" the table by repeatedly using the following simple fact:

$$\text{if } 2^a = A \text{ and } 2^b = B, \text{ then } 2^{(a+b)/2} = \sqrt{AB}.$$

Using this, we get estimates such as

$$2^{0.5} \approx 1.41421, \quad 2^{1.5} \approx 2.82843,$$
$$2^{2.5} \approx 5.65685, \quad 2^{3.5} \approx 11.3137,$$

and so on. These give us facts such as

$$\log_2 1.41421 \approx 0.5, \quad \log_2 2.82843 \approx 1.5, \quad \ldots.$$

By using these newly-computed estimates, we get further estimates:

$$2^{0.25} \approx 1.18921, \quad 2^{0.75} \approx 1.68179,$$
$$2^{1.25} \approx 2.37841, \quad 2^{1.75} \approx 3.36359,$$

and so on. These give us facts such as

$$\log_2 1.18921 \approx 0.25, \quad \log_2 1.68179 \approx 0.75, \quad \ldots.$$

And so we proceed. By-and-by we obtain a fine net of values, with a few gaps in between. Once we have obtained a large number of such estimates, we plot a graph of $\log_2 x$ against x. As the points are placed upon the graph they start to form a curve in a natural and inevitable manner. Using this curve we may now read the value of $\log_2 x$ for any positive number x. The curve is displayed in Figure 4.2.

The arrows show the estimation of $\log_2 12$: we draw a line vertically up from $(12, 0)$ till it hits the curve, then to the left till we hit the vertical axis. We find that $\log_2 12 \approx 3.585$.

4.6 Common logarithms

Logarithms to base 10 are called *common logarithms*. As noted earlier, it was Briggs who persuaded Napier to switch to base 10 for his system of logarithms. The motivation behind the change is not hard to see; after all, it is the powers of 10 with which we are most familiar; we "think" in powers of 10. Observe that

$$\log_{10} 0.01 = -2, \ \log_{10} 0.1 = -1, \ \log_{10} 1 = 0,$$
$$\log_{10} 10 = 1, \ \log_{10} 100 = 2, \ \log_{10} 1000 = 3,$$

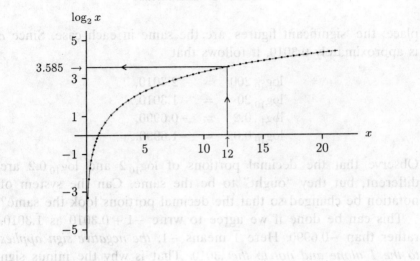

Figure 4.2. *Graph of* $\log_2 x$

and so on; more generally, $\log_{10} 10^n = n$ for all integers n. It should be clear that if we have at hand a detailed list of powers of 10 – i.e., a table of common logarithms – then computations involving multiplication, division and extraction of powers and roots, no matter how complicated, would be routine. Such tables were prepared by Briggs himself. It is staggering to imagine the labour that went into the preparation of the tables. As the quote from Cajori [page 37] suggests, the tables proved to be of enormous help to people who routinely had to perform complex computations. In particular, they "doubled the life of the astronomer".

One of the features about tables of common logarithms that causes confusion is the manner in which negative logarithms are treated. To see how this comes about, consider the relationships between the following quantities:

$$\log_{10} 200, \ \ \log_{10} 20, \ \ \log_{10} 2, \ \ \log_{10} 0.2, \ \ \log_{10} 0.02, \ \ \dots.$$

Write c for $\log_{10} 2$; then

$$
\begin{aligned}
\log_{10} 200 &= 2 + c, \\
\log_{10} 20 &= 1 + c, \\
\log_{10} 0.2 &= -1 + c, \\
\log_{10} 0.02 &= -2 + c,
\end{aligned}
$$

and so on. Observe that 'c' is common to all of them, which is as it should be, as the numbers differ only in the decimal place; the 'significant figures' are the same in each case. Since c is approximately 0.3010, it follows that

$$\begin{aligned}
\log_{10} 200 &= 2.3010, \\
\log_{10} 20 &= 1.3010, \\
\log_{10} 0.2 &= -0.6990, \\
\log_{10} 0.02 &= -1.6990.
\end{aligned}$$

Observe that the decimal portions of $\log_{10} 2$ and $\log_{10} 0.2$ are different, but they "ought" to be the same. Can the system of notation be changed so that the decimal portions look the same?

This can be done if we agree to write $-1 + 0.3010$ as $\bar{1}.3010$, rather than -0.6990. Here $\bar{1}$ means -1; *the negative sign applies to the 1 alone and not to the* .3010. That is why the minus sign has gone "on top". Similarly, we write $\log_{10} 0.2$ as $\bar{1}.3010$, and $\log_{10} 0.02$ as $\bar{2}.3010$, and so on. So in this system, numbers that differ only in the placement of the decimal point have logarithms with the same decimal portion.

The decimal portion of the logarithm of a number is called the *mantissa*, and the integer part is called the *characteristic* of the logarithm. (The nomenclature comes from Napier and Briggs.) That is, if a positive number x is written as $y \times 10^n$ where $1 \le y < 10$ and n is an integer (note that y and n are uniquely fixed by x), then n is the characteristic of $\log_{10} x$, and $\log_{10} y$ is the mantissa. So the mantissa always lies between 0 and 1, and the characteristic is always an integer (negative, positive or zero).

The notation may look strange, but once the ground rules have been established (that is, the arithmetical laws governing such numbers), then the strangeness recedes. For instance, we have

$$\bar{2} + \bar{3} = \bar{5}, \quad \bar{3} + \bar{4} = \bar{7}, \quad 2 \times \bar{3} = \bar{6},$$

and so on. Similarly,

$$\bar{2} - \bar{3} = 1, \quad \bar{3} - \bar{2} = \bar{1}, \quad \bar{7} - \bar{9} = 2, \quad \dots.$$

Division has to be done with care. For example, we have

$$\begin{aligned}
\bar{4} \div 2 &= \bar{2}, & \bar{6} \div 2 &= \bar{3}, & \bar{9} \div 3 &= \bar{3}, & \dots, \\
\bar{3} \div 2 &= \bar{2}.5, & \bar{5} \div 2 &= \bar{3}.5, & \bar{8} \div 5 &= \bar{2}.4, & \dots.
\end{aligned}$$

To see why $\bar{3} \div 2 = \bar{2}.5$, observe that $\bar{3} = \bar{4} + 1$, so

$$\bar{3} \div 2 = (\bar{4} + 1) \div 2 = \bar{2} + 0.5 = \bar{2}.5,$$

and similarly for the other cases.

★ ★ ★

Let us now see how these ideas are put into practice in an actual computational problem: *Find the fifth root of* 0.23. From the printed tables we get $\log_{10} 2.3 = 0.3617$. It follows that $\log_{10} 0.23 = \bar{1}.3617$. We now have

$$\bar{1}.3617 \div 5 = (\bar{5} + 4.3617) \div 5 = \bar{1}.8723.$$

We then read the antilogarithm of 0.8723 from the tables and find the number 7.453. The characteristic being -1, we fix the decimal point accordingly and obtain

$$\sqrt[5]{0.23} = 7.453 \times 10^{-1} = 0.7453.$$

Much more can be said about computational techniques using logarithms but we shall not do so here, as the advent – in recent times – of high-speed electronic calculators has rendered such methods obsolete.

★ ★ ★

Used in conjunction with the usual base-10 system of representing numbers, the idea leads to a non-standard representation system, one where $2\bar{3}$ would means $20 - 3 = 17$, $4\bar{3}$ would mean $40 - 3 = 37$, and $\overline{23}$ would mean -23. The idea can be put to use in doing subtraction:

$$53 - 37 = 53 + \overline{37} = 2\bar{4} = 20 - 4 = 16,$$

and so on. Children need to become familiar with such systems, if only because they show that representation systems are not unique, with each system possessing its merits and demerits. Also, the system is good for providing manipulative exercises.

EXERCISES

4.6.1 Given that $\log_{10} 2 = 0.3010$ and $\log_{10} 3 = 0.4771$, find the logarithms to base 10 of the following numbers: 12, 120, 576 and 48.

4.6.2 Prepare a table of values of powers of 3, as was done for the powers of 2 (above), and find $\log_3 x$ for as many values of x as possible. Then draw a graph of $\log_3 x$.

4.6.3   Write c for $\log_2 3$. Show that

(a) $\log_{16} 24 = \dfrac{c+3}{4}$,

(b) $\log_{12} 24 = \dfrac{c+3}{c+2}$, and

(c)   find a similar expression for $\log_{18} 36$.

4.6.4   Let $\log_{20} 80$ be denoted by a.

(a)   Show that $\log_2 5 = \dfrac{4-2a}{a-1}$.

(b)   Express $\log_{10} 50$ in terms of a.

Chapter 5

Computation of Logarithms

In this chapter, we show how logarithms can be computed to high levels of accuracy, using systematic and precise algorithms which are very different in spirit from the rather *ad hoc* 'close encounter' procedures used in the preceding chapter. The chapter is more advanced than the others in the book, and younger readers may want to skip lightly through it, or even omit it altogether.

5.1 Logarithms in binary form

We start by presenting an iterative[1] technique for computing $\log_{10} x$ in base 2 (i.e., in the binary system), for any given x with $1 < x < 10$. The output of the procedure is a string of 0s and 1s, which, with a decimal point placed before it, is the value of $\log_{10} x$ in binary notation. At first sight the method looks bizarre, but it is easy to understand once the details are clear. Here is the procedure in "pseudo-code form".

Step 0. (Initialization) Input x $(1 < x < 10)$; let $z = x$.

Step 1. Compute $y = z^2$. If $y < 10$ print '0', else print '1'.

Step 2. If $y < 10$ then replace z by y, else replace z by $y/10$. Go back to Step 1.

---

[1]A procedure is described as 'iterative' when it consists of some action repeated over and over again.

47

Observe that we have not provided an "escape route", so the procedure will go on forever (it will be always stuck in the following loop: Step 1, Step 2, Step 1, Step 2, ...). It is easy, however, to provide an escape route, in the form of a stopping condition that comes into force once a sufficient number of digits have been computed.

Let us see what the procedure gives when $x = 2$. The successive computations are displayed below.

Step 1: $z = 2$, $y = 4 < 10$, $\therefore$ output is 0.
Step 2: $y < 10$; $\therefore$ replace z by $y = 4$, repeat step 1.
Step 1: $z = 4$, $y = 16 > 10$, $\therefore$ output is 1.
Step 2: $y > 10$; $\therefore$ replace z by $y/10 = 1.6$, repeat step 1.
Step 1: $z = 1.6$, $y = 2.56 < 10$, $\therefore$ output is 0.
Step 2: $y < 10$, $\therefore$ replace z by $y = 2.56$, repeat step 1.

And so on .... After continuing for 19 iterations, we get the z-values shown below, rounded off to 4 decimal places. (The starting value too has been given, namely $z = 2$.)

$$2.0, \quad 4.0, \quad 1.6, \quad 2.56, \quad 6.5536,$$
$$4.295, \quad 1.8447, \quad 3.4028, \quad 1.1579, \quad 1.3408,$$
$$1.7977, \quad 3.2317, \quad 1.0444, \quad 1.0907, \quad 1.1897,$$
$$1.4155, \quad 2.0035, \quad 4.0141, \quad 1.6113, \quad 2.5964.$$

The next z-value is 6.7411. The corresponding output digits (0 or 1) are, therefore,

$$0, 1, 0, 0, 1, 1, 0, 1, 0, 0, 0, 1, 0, 0, 0, 0, 0, 1, 0, 0, 1,$$

and it follows that $\log_{10} 2$ is given in binary form by

$$\log_{10} 2 = 0.01001101000100000 1001\ldots.$$

Translating this into fractions, we obtain

$$\log_{10} 2 = \frac{1}{4} + \frac{1}{32} + \frac{1}{64} + \frac{1}{256} + \frac{1}{4096} + \cdots.$$

The first 20 terms yield $78913/262144$ or approximately 0.301029. The actual value is $0.3010299956\ldots$ or roughly 0.30103; note how close the two values are.

$$\star \star \star$$

Consider next the case $x = 3$. The z-values obtained from the first 9 iterations are shown below, together with the first value ($z = 3$).

$$3.0, \quad 9.0, \quad 8.1, \quad 6.561, \quad 4.3047,$$
$$1.853, \quad 3.4337, \quad 1.179, \quad 1.39, \quad 1.9323.$$

The next value (the 10th) is 3.7339. The corresponding output binary digits are 0, 1, 1, 1, 1, 0, 1, 0, 0, 0, 1, so it follows that

$$\log_{10} 3 = \frac{1}{4} + \frac{1}{8} + \frac{1}{16} + \frac{1}{32} + \frac{1}{128} + \frac{1}{2048} + \cdots,$$

so $\log_{10} 3 \approx 977/2048 = 0.477051$. For comparison, the true value of $\log_{10} 3$ is 0.477121.

★ ★ ★

Here is a computer implementation, in BASIC, of the above algorithm.

```
REM Finding log x to base 10 in binary form.
CLS
INPUT x : z = x
FOR n = 1 TO 20
    y = z^2
    IF y < 10 THEN PRINT "0" ELSE PRINT "1"
    IF y < 10 THEN z = y ELSE z = y/10
NEXT n
```

EXERCISES

5.1.1 Compute $\log_{10} 5$ and $\log_{10} 7$ in binary form using the technique.

5.1.2 Let x be a number between 1 and 10, and let $\log_{10} x$ be written in binary form as

$$\log_{10} x = 0.a_1 a_2 a_3 a_4 \ldots,$$

with each a_i equal to 0 or 1. Show that $a_1 = 0$ if $x^2 < 10$, or else $a_1 = 1$. Find a similar criterion for fixing the value of a_2.

5.1.3 What is the rationale behind the method?

5.2 Logarithms via square roots

The algorithm considered in this section is the one originally used by Briggs. In essence, it is the same as the algorithm presented in the preceding section, but the implementation is very different.

The basic idea is simple. Let x be a positive number, and suppose that numbers $a, b, c, \ldots$ can be found such that

$$x = 10^a \times 10^b \times 10^c \times \cdots;$$

then, clearly, $\log_{10} x = a + b + c + \cdots$. The algorithm described below shows how the idea is put into practice.

We start by computing the quantities $10^{1/2}$, $10^{1/4}$, $10^{1/8}$, $10^{1/16}$, $\ldots$. The table displayed below has been prepared by successive extraction of square roots. For convenience, we write a_i for $10^{1/2^i}$ (with $a_0 = 10$ by definition).

$$a_1 = 3.162278, \quad a_2 = 1.778279, \quad a_3 = 1.333521,$$
$$a_4 = 1.154782, \quad a_5 = 1.074608, \quad a_6 = 1.036633,$$
$$a_7 = 1.018152, \quad a_8 = 1.009035, \quad a_9 = 1.004507,$$
$$a_{10} = 1.002251, \quad a_{11} = 1.001125, \quad a_{12} = 1.000562,$$
$$a_{13} = 1.000281, \quad a_{14} = 1.000141, \quad a_{15} = 1.000070,$$
$$a_{16} = 1.000035, \quad a_{17} = 1.000018, \quad a_{18} = 1.000009,$$

and so on. (Note that the figures are given accurate to 6 decimal places; Briggs used more than 15 decimal places!) We now show how the table can be used to compute $\log_{10} x$ for $1 < x < 10$. The following algorithm is used.

Step 1. (Initialization) Input x $(1 < x < 10)$; let $z = x$, $L = 0$.

Step 2. Find i such that $a_i \leq z < a_{i-1}$; i.e., such that

$$10^{1/2^i} \leq z < 10^{1/2^{i-1}}.$$

Step 3. (Updating) Update z and L as follows:

$$z \to z \div a_i; \quad L \to L + 2^{-i},$$

with i as found in Step 2, above.

Step 4. (Criterion for Termination) If z differs from 1 by less than 10^{-8}, then stop; else return to Step 2. The estimated value of $\log_{10} x$ is L.

If greater accuracy is desired, we replace the figure of 10^{-8} by a smaller number, say 10^{-12}.

We illustrate the algorithm with $x = 2.5$. At the beginning, we have $z = 2.5$ and $L = 0$.

We start by finding i such that $a_i \leq z < a_{i-1}$. The table gives $i = 2$. We compute z/a_2 and obtain 1.40585. We now have $z = 1.40585$ and $L = 0 + 1/4 = 1/4$.

Once again we seek i such that $a_i \leq z < a_{i-1}$; the table gives $i = 3$. We compute z/a_3 and obtain 1.05424. We now have $z = 1.05424$ and $L = 1/4 + 1/8 = 3/8$.

Continuing, we seek i such that $a_i \leq z < a_{i-1}$; the table gives $i = 6$. We compute z/a_6 and obtain 1.01699. We now have $z = 1.01699$ and $L = 3/8 + 1/64 = 25/64$.

For the next round the table gives $i = 8$. We compute z/a_8 and obtain 1.00788, so we now have $z = 1.00788$ and $L = 25/64 + 1/256 = 101/256$.

And so we proceed . . . . . After 10 rounds, we arrive at $L = \dfrac{13039}{32768}$. So our estimate for $\log_{10} 2.5$ is

$$\log_{10} 2.5 \approx \frac{13039}{32768} \approx 0.39792.$$

For comparison, a more accurate value is 0.39794.

★ ★ ★

With $x = 2$ we obtain, after 26 rounds, $L = \dfrac{20201781}{67108864}$. Our estimate is, therefore,

$$\log_{10} 2 \approx \frac{20201781}{67108864} \approx 0.301029995.$$

For comparison, a more accurate value is 0.301029996.

★ ★ ★

For $x = 3$, we obtain, after 21 rounds, $L = \dfrac{1000595}{2097152}$. So our estimate is

$$\log_{10} 3 \approx \frac{1000595}{2097152} \approx 0.477120876.$$

For comparison, a more accurate value is 0.477121255.

★ ★ ★

A Mathematica implementation of the algorithm is given below (lg(x) refers to the estimate for $\log_{10} x$).

```
f[n_]:= f[n] = N[10^(1/2^n), 8]
g[x_]:= (i = 1; While[f[i] > x, i=i+1]; i)
lg[x_]:= (z = x; s = 0; Do[(s = s + 1/2^g[z];
         z = z/f[g[z]]), {10}]; s)
```

EXERCISES

5.2.1 Compute $\log_{10} 3$ and $\log_{10} 7$ using Briggs' method.

5.2.2 What needs to be done if $x < 1$ or $x > 10$?

5.3 Interpolation techniques

The technique described below is once again based on the repeated extraction of square roots, but it uses *interpolation*—a principle of very great versatility in numerical methods.

The underlying idea, which we use repeatedly, is this: *If a and b are positive numbers which are very close to one another, then their arithmetic mean (AM) and geometric mean (GM) are very nearly equal.* That is,

$$\frac{a+b}{2} \approx \sqrt{ab}.$$

A few examples will serve to emphasize the point. If $a = 1$ and $b = 1.02$, then the AM is 1.01 and the GM is 1.00995; note the closeness of the two numbers. For $a = 20$ and $b = 20.1$, we obtain 20.05 for the AM and 20.0499 for the GM. The important point to be observed is that the AM and GM are very much closer than the original pair of numbers.

Let a and b be two positive numbers which are close to one another. Then, as already noted, $(a+b)/2$ and $\sqrt{ab}$ are very nearly equal to one another. Now, the following relation holds exactly: for any two positive numbers a, b,

$$\log_{10} \sqrt{ab} = \frac{\log_{10} a + \log_{10} b}{2}. \tag{3.1}$$

Combining (1) with the earlier observation, we deduce that if a and b are positive numbers which are very close to one another, then

$$\log_{10} \frac{a+b}{2} \approx \frac{\log_{10} a + \log_{10} b}{2}. \tag{3.2}$$

Relations (1) and (2), together with the observation that for any whole number n,

$$\log_{10} 10^n = n \quad \text{and} \quad \log_{10} 10^{1/2^n} = 1/2^n, \qquad (3.3)$$

will serve as the basis of our technique to compute a table of common logarithms.

We start by using (1) and (3) to compute the logarithms of as many numbers as possible. Since $10^{1/2^n}$ can be computed by hand, using the old-fashioned "long-division" method for extraction of square roots, this can be done without too much trouble. In this manner, we precisely compute the common logarithms of a large number of numbers of the form $10^{m/n}$, where m is a whole number and n is a power of 2. We obtain results such as the following:

- $10^{1/2} = 3.16228$, so $\log_{10} 3.16228 = 0.5$;

- $10^{1/4} = 1.77828$, so $\log_{10} 1.77828 = 0.25$;

- $10^{1/8} = 1.33352$, so $\log_{10} 1.33352 = 0.125$;

- $10^{1/16} = 1.15478$, so $\log_{10} 1.15478 = 0.0625$;

- $10^{3/4} = 5.62341$ (note that $10^{3/4}$ is the square root of the square root of 10^3), so $\log_{10} 5.62341 = 0.75$;

- $10^{7/8} = 7.49894$ (note that $10^{7/8}$ is the square root of the square root of the square root of 10^7), therefore $\log_{10} 7.49894 = 0.875$;

and so on. The idea now is to continue this till we have a fine net of numbers lying between 1 and 10, whose common logarithms we know exactly. This is clearly possible, though the work involved is tedious.

Following this, we use (2) to find the approximate values of the logarithms of the in-between numbers. This technique is known as *interpolation*. We shall not carry out the computations here, but the details of the method should be clear.

EXERCISES

5.3.1 Compute $10^{3/8}$ using repeated extraction of square roots, and use the results to add two more numbers to the list shown above.

5.3.2 Do the same for $10^{5/16}$ and $10^{5/32}$.

5.3.3 Estimate the value of $\log_{10} 6$.

5.4 Interlude on interpolation

Interpolation methods are very widely used in mathematics. The underlying idea is simple. Let f be an "nice" function. By this, we mean that f does not show wild changes in behaviour; its graph has no points of discontinuity (i.e., the graph is in one piece), and it is 'smooth' (i.e., it has no kinks). Then for any two sufficiently close numbers a and b in the domain of f, the following approximate relation holds:

$$f\left(\frac{a+b}{2}\right) \approx \frac{f(a)+f(b)}{2}.$$

We consider a few examples to illustrate this statement.

- Let f be the square root function, $f(x) = \sqrt{x}$, defined for $x > 0$, and let $a = 25$ and $b = 25.1$. Then $\sqrt{a} = 5$, $\sqrt{b} = 5.00999$, $(a+b)/2 = 25.05$ and

$$\sqrt{(a+b)/2} = 5.0049975, \qquad \frac{\sqrt{a}+\sqrt{b}}{2} = 5.004995.$$

  The two values match almost exactly!

- Consider the squaring function, $f(x) = x^2$. Let $a = 5$, $b = 5.1$. Then $a^2 = 25$, $b^2 = 26.01$, $(a+b)/2 = 5.05$ and

$$\left(\frac{a+b}{2}\right)^2 = 25.5025, \qquad \frac{a^2+b^2}{2} = 25.505.$$

  The closeness is less impressive in this case (though this could have been anticipated; do you see why?).

- Consider the reciprocal function, $f(x) = 1/x$. Let $a = 5$, $b = 5.1$; then $1/a = 0.2$, $1/b = 0.196078$, $(a+b)/2 = 5.05$, and

$$\frac{1}{(a+b)/2} = 0.19802, \qquad \frac{1/a+1/b}{2} = 0.198039.$$

  The closeness is more apparent here.

- Consider the sine function, $f(x) = \sin x$. Let $a = 30°$ and $b = 31°$. We have, $\sin 30° = 0.5$, $\sin 31° = 0.515038$, $(a+b)/2 = 30.5°$, and

$$\sin \frac{a+b}{2} = 0.507538, \qquad \frac{\sin a + \sin b}{2} = 0.507519.$$

There is agreement to four decimal places.

The functions for which interpolation works exactly are the *linear* functions; i.e., the functions f for which $f(x)$ is of the form $ax + b$, for some pair of constants a and b. For this reason, the search for linear approximations to functions is a vitally important first step in understanding their behaviour. It is also, therefore, a useful tool in solving practical problems (e.g., in engineering applications).

5.5 Using the exponential function

The algorithm that we present here differs substantially from those described in the preceding sections (which have a certain commonality, being based on the extraction of square roots). It uses more advanced concepts and is perhaps correspondingly less accessible to the younger reader. (More will be said on this matter in Chapter 11.)

We use the following facts:

(a) If h is close to 0 (this is sometimes written as $|h| \ll 1$, or $|h| \approx 0$) and n is a positive integer (the larger the better), then

$$\sqrt[n]{1 + h} \approx 1 + \frac{h}{n}.$$

EXAMPLE. If $h = 0.003$ and $n = 3$, this gives $\sqrt[3]{1.003} \approx 1.001$; the true value of $\sqrt[3]{1.003}$, accurate to 10 d.p., is 1.000999002. And if $h = 0.002$ and $n = 4$ this gives $\sqrt[4]{1.002} \approx 1.0005$; the true value, accurate to 10 d.p., is 1.00499625.

A much closer approximation is

$$\sqrt[n]{1 + h} \approx 1 + \frac{h}{n} - \frac{(n-1)h^2}{2n^2},$$

which gives $\sqrt[3]{1.003} \approx 1.000999$.

These results come from the Binomial Theorem, first enunciated by Isaac Newton.

Chapters 10 and 11 have a lot more information on e and also on the exponential function.

(b) There is a number known as *Euler's number*, denoted by the letter e (its value is roughly 2.71828), with the special property[2] that for all sufficiently small values of h (i.e., for $|h| \ll 1$), we have the approximate relation

$$e^h \approx 1 + h.$$

It is precisely this simple relation which makes e so special, and so ubiquitous in its occurrence in the natural sciences (in physics, chemistry and biology).

A closer approximation is: $e^h \approx 1 + h + h^2/2$, and a still closer one is: $e^h \approx 1 + h + h^2/2 + h^3/6$.

(c) Logarithms to base e are called *natural logarithms*, and $\log_e x$ is generally denoted by the symbol $\ln x$; the 'ln' stands for 'natural logarithm'. The relation quoted in (b) shows that if $|h| \ll 1$, then $\ln(1 + h) \approx h$. A closer approximation is

$$\ln(1 + h) \approx h - \frac{h^2}{2},$$

and closer still is: $\ln(1 + h) \approx h - h^2/2 + h^3/3$.

As a corollary we get: if $|h| \ll 1$, then $\ln(1 - h) \approx -h - h^2 - h^3/3$, and so, by subtraction,

$$\ln \frac{1 + h}{1 - h} \approx 2 \left(h + \frac{h^3}{3} \right).$$

Closer still is: $\ln \dfrac{1 + h}{1 - h} \approx 2 \left(h + \dfrac{h^3}{3} + \dfrac{h^5}{5} \right).$

(d) Using special techniques, the value of $\ln 10$ can be computed to high levels of accuracy; we find that

$$\ln 10 \approx 2.3026, \quad \log_{10} e \approx 0.4343.$$

More accurate values are

$$\ln 10 = 2.302585092, \quad \log_{10} e = 0.4342944819.$$

---

[2]The number e is defined to be the sum of the infinite series: $1 + 1/1 + 1/2 + 1/6 + 1/24 + 1/120 + \cdots$, the denominators of the fractions being the *factorial numbers*: $2 = 1 \cdot 2$, $6 = 1 \cdot 2 \cdot 3$, $24 = 1 \cdot 2 \cdot 3 \cdot 4$, and so on. Its value to 9 decimal places is 2.718281828 and it is an example of an irrational number—it cannot be written as a ratio of two whole numbers. It has a way of turning up in all sorts of contexts. For example, it comes up in the study of prime numbers!

(e) From (b) and (d), we conclude that if $|h| \ll 1$, then

$$\log_{10}(1 + h) \approx 0.4343h.$$

A closer approximation is given by

$$\log_{10}(1 + h) \approx 0.4343 \left(h - \frac{h^2}{2} \right).$$

(f) From (a) and (e), it follows that if

$$a^n \approx (1 + h) \times 10^m,$$

where n is "reasonably large" and h is as close to 0 as possible
($|h| \ll 1$; h is allowed to be negative), then

$$\log_{10} a \approx \frac{m + (0.4343 \times h)}{n}.$$

A closer approximation is

$$\log_{10} a \approx 0.4343 \left(\frac{h}{n} - \frac{(n-1)h^2}{2n^2} \right) + \frac{m}{n}.$$

$$\star \; \star \; \star$$

We now show how these facts aid the computation of logarithms.
We shall begin with the computation of $\log_{10} 2$. Our starting point
is the same as earlier—the observation that $2^{10} = 1024$ is close to
a power of 10. From this we see, by taking 10th roots, that

$$2 = \sqrt[10]{1.024} \times 10^{0.3}.$$

Using (a), above, we find that $\sqrt[10]{1.024} \approx 1.0024$, so $2 \approx 1.0024 \times 10^{0.3}$. Taking logarithms on both sides,

$$\log_{10} 2 \approx \log_{10} 1.0024 + 0.3.$$

Using (d), we find that

$$\log_{10} 1.0024 \approx 0.4343 \times .0024 \approx 0.001.$$

Putting everything back together again, we see that

$$\log_{10} 2 \approx 0.301.$$

Comparing this with the value given earlier, we can see that we have achieved a fair degree of accuracy.

We can do better; a closer estimate of $\sqrt[10]{1.024}$ is

$$1 + .0024 - \frac{9}{100} \frac{.024^2}{2} \approx 1.00237.$$

Next,

$$\log_{10} 1.00237 \approx .4343 \left(0.00237 - \frac{.00237^2}{2} \right) \approx 0.00103.$$

It follows that

$$\log_{10} 2 \approx 0.00103 + 0.3 = 0.30103,$$

and this is an excellent approximation.

★ ★ ★

Let us see what the relation $3^2 = 9 = 0.9 \times 10$ tells us about the value of $\log_{10} 3$. Rewriting the relation as $3 = \sqrt{1 - 0.1} \times 10^{0.5}$ and comparing with (f), above, we see that $n = 2$, $m = 1$ and $h = -0.1$. Therefore,

$$\log_{10} 3 \approx 0.4343 \left(\frac{-0.1}{2} - \frac{0.01}{8} \right) + \frac{1}{2},$$

or $\log_{10} 3 \approx 0.4788$. This is not too far from the true value of 0.4771 (but not too close either!). The crudity of the approximation owes to the fact that $n = 2$ is not large enough and $h = -0.1$ not small enough.

For better results, we use the result quoted in (e) and proceed thus: $\sqrt{1 - 0.1} \approx 0.9487 = 1 - 0.0513$, and

$$\log_{10}(1 - 0.0513) \approx 0.4343 \left(-0.0513 - \frac{0.002632}{2} \right)$$
$$\approx -0.4343 \times 0.0526 \approx -0.0228.$$

Therefore, $\log_{10} 3 \approx 0.5 - 0.0228 = 0.4772$: almost on the dot, because $\log_{10} 3 = 0.477121\ldots$.

We can also make use of earlier findings, rather than opting to start afresh each time. As an example, we show how this can be done for the estimation of $\log_{10} 7$. We use the relation $7^2 = 49$:

$$2 \times 7^2 = 98 = (1 - 0.02) \times 10^2.$$

Taking logarithms of both sides, we obtain

$$\log_{10} 2 + 2\log_{10} 7 = \log_{10}(1 - 0.02) + 2.$$

Next, we estimate the value of $\log_{10}(1 - 0.02)$:

$$\log_{10}(1 - 0.02) \approx 0.4343 \left(-0.02 - \frac{0.02^2}{2} \right) \approx -0.00878,$$

giving $\log_{10} 2 + 2\log_{10} 7 \approx 1.99122$. Since $\log_{10} 2 \approx 0.30103$, this yields

$$\log_{10} 7 \approx \frac{1.99122 - 0.30103}{2} \approx 0.845095.$$

In fact, $\log_{10} 7 = 0.845098\ldots$; note the closeness of our estimate.

EXERCISES

5.5.1 Compute the values of $\log_{10} 11$ and $\log_{10} 13$.

5.5.2 Show that $\ln 2 \approx 0.693147$.

Comments on Chapters 6–9

In the following four chapters, we present a number of areas of application of the log function in science and commerce. The reason why in these applications the log function has been preferred to other contenders in the field is not the same in the various examples, but clearly it has something to do with the special properties of the log function. Two of these properties have been listed below.

Bringing down the mighty One of the useful features which the log function has is the ability to "bring down the mighty"—to deflate very large numbers in such a way that their magnitude can be grasped. For example, consider the number 10^{100}. This is a gigantic number! However, its logarithm to base 10 is quite small, merely 100, and another application of the log function yields 2, a pygmy! This illustrates the point.

This property comes of use in applications in astronomy; in the decibel scale for measuring sound; in the pH scale for describing the degree of acidity or alkalinity of a solution; in the Richter scale for describing the intensity of an earthquake; and so on.

Decreasing rate of growth An extremely important property possessed by the log function has to do with its rate of change: *The rate of increase of* $\log x$ *with* x *steadily decreases with* x. Thus, for all $x > 0$, we have the inequality

$$\log(x+2) - \log(x+1) < \log(x+1) - \log x.$$

EXAMPLE. We have, with $x = 5$,

$$\log 6 - \log 5 = 0.07918, \quad \log 5 - \log 4 = 0.09691,$$

and, of course, $0.07918 < 0.09691$. The reader should check other such relations; e.g., $\log 101 - \log 100 < \log 100 - \log 99$.

It is not hard to see why the relationship holds: the quantity on the left is

$$\log(x+2) - \log(x+1) = \log \frac{x+2}{x+1},$$

while that on the right is

$$\log(x+1) - \log x = \log \frac{x+1}{x}.$$

To compare these two quantities, we use the inequality

$$x(x+2) < (x+1)^2,$$

which may be checked via simple multiplication. The inequality now yields, after transposition of terms,

$$x(x+2) < (x+1)^2,$$

$$\therefore \quad \frac{x+2}{x+1} < \frac{x+1}{x},$$

$$\therefore \quad \log\frac{x+2}{x+1} < \log\frac{x+1}{x},$$

$$\therefore \quad \log(x+2) - \log(x+1) < \log(x+1) - \log x.$$

This relationship is of great importance and lies at the basis of many applications.

Conversion of multiplication into addition   The most basic property of the logarithmic function is, of course, that it converts a multiplicative relationship into an additive one.

★ ★ ★

These, then, are some properties of the log function that endear it so greatly to nature and to students of nature and mathematics. The student should look back at this list after reading Chapters 6–9.

Chapter 6

Compound Interest

6.1 Interest

Everyone needs to know about Compound Interest, if only because the loans we take and the way banks operate our savings accounts depend directly upon the system of Compound Interest. Let P be the principal (i.e., the sum of money put into the bank at the start) and let r be the annual rate of interest (expressed as a decimal fraction; e.g., if the rate of interest is 5%, then $r = 0.05$; and so on). The interest in the first year is Pr, so the amount at the end of the first year is

$$P + Pr = P(1 + r).$$

This, in effect, becomes the principal for the second year. Since the ratio of amount to principal is the same each year, namely $(1 + r) : 1$, the amount at the end of the second year is $P(1 + r) \times (1 + r) = P(1 + r)^2$. Similarly, the amount at the end of the third year is $P(1 + r)^2 \times (1 + r) = (1 + r)^3$. The argument is easy to carry forwards, and we find that the amount at the end of the nth year is $P(1 + r)^n$.

$$\star \; \star \; \star$$

We may contrast this formula with that obtained if the interest each year is applied only on the original amount deposited, i.e., on the principal. In this case the interest in each year is the same, namely Pr, so after n years the amount is $P(1 + nr)$. This is generally termed as *simple interest*.

$$\star \; \star \; \star$$

Compounding can also be done *half-yearly*; the unit of time now is half a year and the effective rate of interest is $\frac{1}{2}r$ (per the modified unit of time). In this case, the amount at the end of the nth year will be $P(1+\frac{1}{2}r)^{2n}$, as the number of units of time is $2n$.

One can similarly consider the possibility of the interest being applied three times a year, at equal intervals of time; or four times a year; and so on. Clearly, the more frequently compounding is done the more favourable it is to the investor. If compounding is done k times a year, then the amount at the end of the nth year will be

$$P\left(1+\frac{r}{k}\right)^{nk},$$

the number of units of time being kn.

EXAMPLE. Let the principal be Rs 100/- and let the rate of interest be 12%. If compounding is done annually, then the amount at the end of the first year is Rs $(1+0.12)\times 100 = 112.00$. If compounding is done half-yearly, then the amount at the end of the first year is Rs $(1+0.06)^2 \times 100 = 112.36$; and if it is done thrice-yearly, then the amount at the end of the year is Rs $(1+.04)^3 \times 100 = 112.49$. If compounding is done 12 times a year (i.e., on a monthly basis), then the amount at the end of the year is Rs $(1+.01)^{12} \times 100 = 112.68$. The reader will note the increased benefit obtained by having the compounding done more frequently.

However, there is a limit to the benefit obtained by having the compounding done more and more frequently in this way; but more of that later.

6.2 Doubling your money

A question that most investors would like answered is: *How long will my money take to double?* A very useful thumb rule is the following:

If the interest rate is $i\%$, then the number of years needed for the money to double is roughly $70/i$ years.

This rule gives a close approximation to the true answer when i is not too large. Let us see why the rule works.

Let n be the number of years taken for the money to double. If i is the rate of interest (expressed as a percentage), then $r = i/100$

and we have the relation $P(1+r)^n = 2P$, which yields the following equation to be solved for n:

$$(1 + r)^n = 2.$$

Taking logarithms on both sides, we obtain

$$n \log_{10}(1 + r) = \log_{10} 2.$$

Now, $\log_{10} 2 \approx 0.3010$, and we know from an earlier chapter (Section 5.5) that for small r, $\log_{10}(1 + r) \approx 0.4343\, r$. So we get,

$$n(0.4343\, r) \approx 0.3010, \quad \therefore \quad n \approx \frac{0.3010}{0.4343\, r}.$$

Since $0.3010/0.4343 = 0.693069 \approx 0.7$, we get $n \approx 0.7/r$. Finally, since $i = r/100$, we obtain $n \approx 70/i$.

$$\star \;\star\; \star$$

It is easy to check the accuracy of the thumb rule. Let the rate of interest be 5%; then $i = 5$, $r = 0.05$. According to the rule, the number of years it takes for the money to double is $70/5$ or 14 years. Since $(1.05)^{14} \approx 1.98$, which is nearly equal to 2, this must be very close to the actual answer. (More accurate computations show that the actual answer is 14.2 years, or roughly 14 years, 2 months.)

If the interest rate is 10%, then the rule gives 7 years for the doubling period. The actual doubling period is 7.27 years or 7 years, 3 months. Observe that as the interest rate grows, the accuracy of the rule decreases. This could have been anticipated, as the relation $\log_{10}(1 + r) \approx 0.4343\, r$ holds only when r is small.

REMARK 1. Since population growth follows the same law as the one which governs the growth of money in a bank, it follows that the "law of doubling" holds for population growth too. So if population grows at a rate of $r\%$ per year, then it will double in roughly $70/r$ years. This means in particular that the doubling time remains constant, century after century—provided, of course, that the rate of growth remains the same.

REMARK 2. For purely practical reasons, and the fact that many people have difficulties with arithmetic, we may opt to replace the formula $70/i$ by $72/i$. The reason for the choice is that 72 is a nice 'round' number, with plenty of divisors. So the formula yields a doubling figure of 36 years for a rate of interest of 2%; 24 years for a rate of 3%; and 12 years for a rate of 6%. The actual figures, for comparison, are 35 years, 23.45 years, and 11.9 years respectively; quite close!

EXERCISES

6.2.1 Find a thumb rule that gives the number of years for the money to triple.

6.2.2 Does population growth really follow the same law as that which governs the growth of money in a bank?

Chapter 7

Sensational Stuff

7.1 Stimulus and sensation

Imagine that you are holding a 10 gm weight in your hand. Let an extra weight of 1 gm be placed in your hand. Do you 'feel' the extra weight? Do you sense that something extra has been placed in your hand? Now, let the 10 gm weight be replaced by a 100 gm weight. Do you 'feel' the presence of an extra weight of 1 gm placed in your hand? It seems reasonable to assume that you are less likely to do so in the second situation; and less likely still if you start out with a 1000 gm weight in the hand. Stated otherwise, we are less likely to detect the difference between 101 gm and 100 gm as compared with the difference between 11 gm and 10 gm; and less likely still to detect the difference between 1001 gm and 1000 gm.

In a like manner, we may say that we are less likely to 'feel' the temperature difference between 25°C and 26°C as compared with that between 10°C and 11°C; and still less likely to detect the difference between 60°C and 61°C.

This suggests that the sensitivity of the senses (that is, the ability of the senses to detect an increase in the level of the stimulus) decreases as the magnitude of the stimulus increases. Ernst Weber (1795–1878) hypothesized that the following law applies to such situations:

> The minimum increase of stimulus which will produce a perceptible increase of sensation is proportional to the pre-existing stimulus.

This looks reasonable (after one has understood the words!), but it refers to quantities which are subjective and so cannot be measured in an unambiguous manner. After all, sensations vary from one person to another, even with identical stimuli. It follows that the law cannot be tested in any really scientific sense and so must be regarded as a 'rule of thumb' rather than an exact law.

In the mid-1800s, with the increasing sophistication and variety of measuring instruments, rapid advances were made in the field of biophysics. Gustav Fechner (1801–1887), who worked in a field later to be known as *Psychophysics*, formulated the following law, which is more or less a consequence of Weber's law (Weber and Fechner did several experiments together on hearing).

The response of the senses varies as the logarithm of the stimulus.

Stated otherwise,

If a stimulus increases in a geometric progression, the sensation resulting from it increases in an arithmetic progression.

Let us first see why the alternative formulation follows from Weber's law. Consider the GP with common ratio 2,

$$1, \ 2, \ 4, \ 8, \ 16, \ 32, \ 64, \ \ldots .$$

Taking logarithms to base 2, we get the numbers

$$0, \ 1, \ 2, \ 3, \ 4, \ 5, \ 6, \ \ldots ,$$

an AP with common difference 1. The same thing happens regardless of the base chosen, as may easily be checked.

We shall refer to the above law as the *Weber–Fechner law*.

The term 'senses' used in the statement of the law includes the sense of sound, sight, smell, and so on. In symbolic language, if Y denotes the sensation produced by a stimulus X, then X and Y are related by a law of the form

$$Y = \log_b kX,$$

for some constants k and b, which depend on the particular situation at hand.

The 'why' of the law is hard to account for, as it involves the functioning of the senses and the brain, but it is easy to give empirical evidence in support of the law. The two sections to follow may be regarded as case studies in the investigation of the law.

COMMENT. Is the Weber-Fechner 'law' really a law? Some scientists do not think so; indeed, they regard it as a piece of nineteenth-century fiction! At the heart of the controversy is the difficulty of having a well-defined system of measurement: how does one measure the response to a stimulus? There may be very different ways of measuring responses. Perhaps it is best to regard the law as a rough-and-ready rule rather than a precise law, such as Newton's law of gravitation.

However, we shall plough ahead regardless of these objections. The following chapters describe several areas of application of the Weber-Fechner law.

But first, an interlude. Below, we shall reproduce some extracts from a delightful and highly readable series of articles by Professor A M Vaidya in *Bona Mathematica*, Volume **6**, Numbers 1–3, titled *The Versatile Logarithm*. I thank Professor Vaidya and the publishers of *Bona* for their kind permission to me to reproduce extracts from the articles.

7.2 'The Versatile Logarithm'–I

In one of his experiments, Weber put a pebble weighing 20 gm in the right hand of a person and a pebble weighing 20.5 gm in the left hand. The person did not know the weights. He was asked as to which pebble was heavier. He said, "Well, both are equally heavy". Then the weight in the left hand was replaced one by one with gradually increasing weights. But only when a pebble weighing 21 gm was placed in his left hand could the man say that the left-hand pebble was heavier. You see, when he had to compare with a weight of 20 gm, he could not notice a difference of less than 1 gm. When a weight of 40 gm was put in the right hand, it was found that he could not notice a weight of less than 2 gm. In general, within certain limits, Weber found that for two weights which differed by less than 5% from each other, people could not notice the difference. In other words, in the matter of weight, the *just noticeable difference* or *jnd* was 5%.

Weber performed similar experiments with other sensations, like light, sound, etc. He found that the *jnd* for light is 2%, for sound it is 10%, for the smell of rubber it is 12%, and for the taste of salt it is 25%. In other words, we are most sensitive to light and least sensitive to taste.

Based on his experiments, Weber proposed a law.

But let us first understand two technical terms. Anything that invades our senses is a stimulus. The light coming from a bulb, or the sound created by a train, or the weight of an object are all stimuli (plural of stimulus). The effect that a stimulus produces on our senses is called 'sensation'.

What the scientist can measure with his instruments is stimulus. Since sensation is something that we experience, the scientist cannot devise any instruments to measure it. However, we can safely say that greater the stimulus, the greater the sensation. But what is the exact relationship between the intensity of a stimulus (as measured by a scientist) and the sensation produced by it? Are they equal?

Certainly not. As stimulus, a weight of 20 gm is different from a weight of 20.5 gm, but both produce the same sensation. Therefore, sensation cannot be equal to stimulus. As there can be no instruments to measure sensation, we must find a formula connecting stimulus and the sensation produced by it. Once this is done, we can measure stimulus by instruments and then calculate sensation by using the formula. The first step in finding the formula was Weber's law that the jnd at any level of stimulus is a constant multiple of the stimulus. If, when the stimulus is s, jnd is Δs, then $\Delta s/s$ is a constant (does not depend on s). For example, in the case of weights, we have the following experimental results.

s (gm)	Δs (gm)	$\Delta s/s$
20	1	1/20
40	2	1/20
60	3	1/20

Only a few years later, Fechner cleverly used Weber's law to find a method of measuring sensation. He observed that if the stimulus is very very small, you would not feel the weight of it at all. If someone puts a feather on your head, you would not feel the weight of it at all. The feather does have weight (a few milligrams), so the stimulus is there, but there is no sensation. Similarly, if you put just a few grains of sugar in your cup of tea, you would not detect any sweetness at all. So the stimulus has to reach a certain level before it can be felt, before sensation begins. This level of the stimulus is called its 'threshold value'. If the stimulus is less than its threshold value, the sensation is zero.

Suppose the threshold value is s_0. When $s = s_0$, we just feel a sensation. If now we begin to gradually increase stimulus, sensation will not begin to increase simultaneously. We shall feel a change only when a *jnd* is added to the stimulus. If the *jnd* at s_0 is Δs_0, then by Weber's law, $\Delta s_0 / s_0$ is constant. Let us write W for this constant. Then,

$$\Delta s_0 = s_0 W.$$

At $s = s_0$ the sensation t is just beginning, so we may take $t = 0$.

The next change in sensation takes place when $s = s_0 + jnd$, i.e., when $s = s_0 + s_0 W = s_0(1 + W)$. At this stage, suppose the sensation is t_1. So when $s = s_1 = s_0(1 + W)$, $t = t_1$.

Again, let us continue to increase s beyond s_1. The next change in sensation takes place when $s = s_1 + jnd$, i.e., when $s = s_2 = s_1 + W s_1 = s_1(1 + W) = s_0(1 + W)^2$. At this stage, suppose the sensation is t_2. Now, the crucial question is: What is the relation between t_1 and t_2? t_1 is the increase in sensation as a result of an increase in stimulus of 1 *jnd*. Similarly, sensation increases by $t_2 - t_1$ again when 1 *jnd* is added again. As a result of extensive experiments, Fechner propounded the law that there is a constant increase in sensation if stimulus is increased by 1 *jnd* (i.e., it does not depend on the original stimulus).

Thus, $t_2 - t_1 = t_1$ or $t_2 = 2t_1$.

If we continue using Fechner's law while the stimulus is being continuously increased, we get the following information.

Stimulus (s)	Sensation (t)
s_0	0
$s_0(1 + W)$	t_1
$s_0(1 + W)^2$	$2t_1$
$s_0(1 + W)^3$	$3t_1$

If we write $1 + W = r$, then the levels of stimuli are

$$s_0, \quad s_0 r, \quad s_0 r^2, \quad s_0 r^3, \quad \ldots,$$

and the corresponding sensations are

$$0, \quad t_1, \quad 2t_1, \quad 3t_1, \quad \ldots.$$

Note that the stimuli form a geometric progression and the corresponding sensations form an arithmetic progression. So the formula connecting sensation t with stimulus s should be such that when s increases in GP, t increases in AP. What relation, what function has this property? Which is that function which converts GP to AP, i.e., which converts multiplication into addition? Oh well, that function is of course our well-known friend, [the] logarithm! If we take logarithms of the terms of a GP,

$$a, \quad ar, \quad ar^2, \quad ar^3, \quad \ldots$$

we get

$$\log a, \quad \log a + \log r, \quad \log a + 2\log r, \quad \log a + 3\log r, \quad \ldots,$$

which is an AP.

So t must be related to $\log s$. From this, Fechner derived the formula

$$t = k \log \frac{s}{s_0},$$

where k is a constant and s_0 is the threshold value of s (observe that when $s = s + 0$, $t = 0$). The constant k depends on the uni of measuring s.

If stimulus s produces sensation $f(s)$, then the formula for $f(s)$ is

$$f(s) = \begin{cases} 0, & \text{if } s \leq s_0, \\ k \log(s/s_0), & \text{if } s > s_0. \end{cases}$$

Note that

$$f(2s) = k \log \left(\frac{2s}{s_0} \right) = k \left(\log 2 + \log \frac{s}{s_0} \right) = f(s) + k \log 2.$$

This shows that if you double the stimulus, sensation is not doubled, it only increases by $k \log 2$.

This also shows that when a stimulus invades our senses, our body only takes in its logarithm and sends this logarithm to the brain to create a sensation. So if I say that our body acts like a log table, would you still think that I am crazy?

Acknowledgment The material in Section 7.2 has been reproduced from *Bona Mathematica*, Volume 6, Numbers 1–3, by kind permission from the publisher (Bhaskaracharya Pratishthana, 56/14, Erandvane, Damle Path, Off Law College Road, Pune–411 004, Maharashtra, India).

Chapter 8
Measuring Sound

Sound is essentially a vibration moving through a medium. The particles of the medium move back and forth ('vibrate') in the direction of the sound wave, so sound is described as being a "longitudinal vibration". Alternatively, a sound wave may be regarded as a sequence of pressure changes moving through a medium. This is so, regardless of what the medium is (gas, liquid or solid). Sounds can be heard underwater too, just as in air. Whales are familiar with this and communicate with one another in very beautiful ways through "whale songs"; recordings have been made of these songs. Dolphins too have a complex system of communication. Indeed, sounds travel faster in water than in the air, and faster still in metals. The rumble of an approaching train can be sensed in the rails long in advance of the train. When a heavy truck passes over a road, the road vibrates and shudders under the heavy load. When a heavy vehicle moves across a bridge, the entire structure vibrates as if in sympathetic response.

The Tacoma Narrows disaster There have been instances when bridges have responded with such enthusiasm that they have been torn apart! A particularly famous instance is the collapse of the Tacoma Narrows suspension bridge across Puget Sound off the coast of California, in the November of 1940; here the vibrations were set into motion by a strong wind. As the wind increased, the vibrations of the bridge increased correspondingly until finally it was wrenched apart.

It is for this reason that when a troop of soldiers marches across a bridge, they are ordered to 'break step' and move across in loose formation.

8.1 How do we hear sounds?

It is interesting to ponder over the remarkable way in which we hear sounds. When air movements strike the ear drum, a delicate, taut membrane situated deep within the ear channel, the drum vibrates in sympathy with the incoming vibration. The vibration is carried into the 'middle ear' and thence through the 'inner ear' into the brain (in the form of electrical impulses) via a finely balanced structure of bone, fluid and nerves. Finally, it is the brain that does the 'hearing'. So what the 'outer ear' really does is to respond to pressure changes, and it does an extraordinary job of it. It is so sensitive that it can detect pressure changes of less than $10^{-7}\%$ (corresponding to a displacement of air particles by less than 1 atomic dimension)!

Indeed, it does more. Picture yourself sitting on a railway platform with your eyes shut. Hundreds of sounds are pouring into your ears, producing a jumbled mass of vibrations, yet we are able to make out *exactly* what is happening on the platform: the calls of vendors, the announcements being made over the loudspeaker, the movements of trains chugging in and pulling out, the conversations among the porters, people bidding one another goodbye, and so on. Or picture yourself sitting in a room filled with people in conversation. As earlier, numerous sounds are entering your ears, yet you are able to disentangle them with seemingly no effort at all and make out what any particular person is saying. We may even be able to latch onto what someone at the other end of the room is saying. It is truly a miracle!

Feynman offers a lovely analogy to bring home the miraculous nature of the feat which the ear–brain system seems to perform so routinely. Imagine that you are in the corner of a swimming pool with your eyes shut. People are jumping in and out at different places, others are swimming across using different strokes, others are floating by silently, and yet others are indulging in horse-play at the end of the pool, and the waves from all these movements come and slap against your face. Your task is to decipher all that is going on in each part of the pool simply from the feel of the waves upon your face. Sounds impossible to do? Yet this is what our ears do routinely, all day long!

8.2 Qualities of sounds

Sounds may differ from one another in different ways: in *frequency* or pitch (a sound may be high pitched or low pitched), in *intensity* (the sound may be loud or soft), and in *timbre*. The timbre of a musical instrument is its distinctive quality—the unique mix of harmonics that make it sound different from all other musical instruments. For instance, the sounds made by a piano and a violin are perceptibly different from one another even when they have identical frequency and loudness. The same can be said with regard to the human voice—each voice has its own distinctive quality, its own timbre. (This may be traced to the particular structuring of bone and flesh in the individual's throat and mouth. Small changes in structure can lead to significant changes in the timbre of the voice.)

We shall be concerned in the following section with how loudness is given a numerical measure. As mentioned above, the range of acoustic intensity ('amplitude', to use the technical term) that the ear can detect is enormous, so a logarithmic scale seems appropriate. It is relevant to quote some numbers here. The power content of sound is measured in *watts*. The wattage rating of the softest sound that we can hear is of the order 4×10^{-6} watts, and that of the loudest sound that we can hear is of the order 10^2 watts. This is an enormous scale—a variation by 8 factors of magnitude!

8.3 The decibel scale

The questions posed earlier can now be posed with reference to hearing. When the ear hears a sound of some given pitch, by how much must the sound increase in loudness before our ears are able to detect the change? As earlier, it seems reasonable to assume that the sensitivity of the ear to changes in loudness decreases with increasing loudness of the sound. Empirically, it has been found that our ears are unable to detect changes of less than 25% in the energy output.

The fact that the least detectable difference is a *percentage* of the existing energy level, rather than some fixed quantity, is of significance. It suggests a natural scale of loudness, in which successive markers correspond to 25% increases in the energy content of the sound. It also brings to mind the logarithmic relationship in Fechner's law. To see why, observe that 10 increases each by 25% result in an increase in energy level by a factor of 1.25^{10} or roughly 10, and 10 more such steps result a further magnification by a factor of 10. So a movement of 10 steps along the scale of loudness really means a 10-fold increase in the energy level of the sound. The conversion of a multiplicative change into an additive one is exactly what a logarithmic scale is all about.

The unit used to measure the loudness or intensity of a sound is the *decibel*; the symbol used is 'dB'. It is also a measure of the energy content of the sound. A difference of 10 dB between the intensities of two sounds means that one sound has 10 times the amount of energy of the other one.

The threshold of audibility is 0 dB; this may be said to be the weakest sound that a normal human ear can hear. At the upper end, we have the threshold of pain at 140 dB; beyond this point the ear does not so much hear as experience pain.

COMMENT. Strictly speaking, the threshold of hearing depends upon the pitch of the sound; the 0 dB figure is for a pitch of 3000 Hz. At lower frequencies, the threshold is significantly greater, as it is at higher frequencies. The threshold is the lowest for sounds at 3000 Hz. The threshold of pain too depends upon the frequency of the sound, but less so than the threshold of hearing.

★ ★ ★

Table 8.1 gives the decibel levels of some familiar sounds. From the table, we see that the energy content of the sound emanating from a rock show is nearly 10^9 times that of the sound of the background rustle of leaves.

8.4 The musical scale

In the previous section we discussed the decibel scale, which measures intensity of sound. In this section we discuss pitch, a more familiar concept. Everyone knows the difference between a high-pitched sound and a low-pitched sound. The following facts are well-known:

- The sound made by plucking a taut string goes up in pitch as the length of the string is shortened (think of a guitar string).

Table 8.1 *Decibel levels of some familiar sounds*

Source	dB
Threshold of hearing (at 3000 Hz)	0
Gentle rustle of leaves	10
Whisper in a room	20
Conversation at 4 m	50
Normal conversation	60
Telephone bell	70
Circular saw	100
Amplified rock band	120
Threshold of pain	140
Jet take-off (close range)	140
Saturn rocket take-off (close range)	200

- The sound made when one pours water into a container gradually, rises in pitch as the water level rises (as the air column above the water has correspondingly less space to vibrate).

- When recorded music is played back at a higher-than-normal speed (say, a 33-rpm record played at 45 rpm or 78 rpm), the music is heard at a higher pitch. Interestingly, this does not make the music go out of pitch, as the frequencies have all gone up by the same ratio and, therefore, continue to bear the same proportions to one another; so the 'tune' of the song does not change—Fechner's law once again!

And so on .... The inference is clear—that an increase in pitch results from particles vibrating at higher speed, which in turn happens when there is less space within which to vibrate. This can be observed with regard to our voices as well: when one is young the vocal chords are short and the voice has a high pitch. As one grows older, the vocal chords lengthen and the voice acquires a lower pitch (the voice "deepens").

Pitch is denoted symbolically on the musical scale, which takes the following forms:

- In the Western system:
  DO–RE–MI–FA–SO–LA–TE–$\overline{\text{DO}}$; or

- In the Indian system:
  SA–RE–GA–MA–PA–DA–NI–$\overline{\text{SA}}$.

The second DO or SA (written $\overline{\text{DO}}$ and $\overline{\text{SA}}$, respectively) is one octave higher than the first. One of the first things learnt in music is that sounds which are 1 octave apart, made by the same instrument, sound almost the same. In particular, when the two sounds are made simultaneously, the ear finds it hard to sense that there are two distinct sounds being made; they blend into a single sound. We have probably all experienced the phenomenon, while singing, of "dropping an octave" (lapsing into the lower octave). On the other hand, when sounds of different pitch are played together and do not differ by an octave (or by a multiple of one octave), the ear can instantly detect that "something is wrong". This is what musical training is supposed to train us to do!

Pitch is measured in cycles per second or *Hertz*, Hz (named after Heinrich Hertz, the physicist who first generated electromagnetic waves in the laboratory). The higher the Hz figure, the higher the pitch. The human ear is sensitive to a fairly wide range of frequencies. Below the lower cut-off point the ear hears nothing. Above the upper end point too the ear hears nothing, but the ear can be damaged if the sound is too 'loud'—it can be loud even though the ear hears nothing! This is 'ultrasound', and it has a wide variety of uses, particularly in medicine.

It was Pythagoras and his contemporaries who first discovered the numerical laws governing pitch. They found, for instance, that the sounds made by strings whose lengths are in the ratio 1 : 2 (other factors remaining constant: the width and density of the string, the tension in the string, ...) differ by exactly one octave. Since halving the length leads to doubling the frequency, this shows that when the frequency doubles, we climb up by 1 octave. So if a certain sound is marked as DO, then $\overline{\text{DO}}$ (the next higher DO) has double the frequency of the first one. This is a remarkable discovery. It shows clearly how Fechner's law operates for pitch. Once this idea has been grasped, the construction of the musical scale becomes an easy matter.

* * *

We reproduce below further extracts from the article by Professor A M Vaidya, which we have already encountered in the previous chapter.

8.5 'The Versatile Logarithm'–II

We saw in the last article that our senses are invaded by stimuli such as sound, light, fragrance, taste, etc. The reaction of our senses to this invasion is the sensation which we experience. The sensation is produced by stimulus but that does not mean that the intensity of sensation is equal to that of the stimulus, or even that it is proportional to it. When stimulus increases, sensation also increases but there is a lot of difference in the rates at which they increase. The sensation increases very slowly and almost in jerks. According to Weber and Fechner, the sensation is related to the logarithm of the stimulus. Because of this, the units to measure sensation have to be very different from the units to measure stimulus.

Let us understand this by an example. Take the sound stimulus. Scientists measure the intensity of sound (stimulus) in units of Watts per square meter. You must have noticed that the sound output of a music stereo is expressed in Watts. It is actually Watts/m^2. If sound is produced by a vibration of 100 Hertz/sec, then its threshold value is 10^{-12} Watts/m^2. That is, if the intensity of the sound is less than this value, then we cannot hear it. We shall denote this threshold value by s_0.

As we saw above, the sensation t produced by a stimulus of s Watts/m^2 is given by

$$t = k \log \frac{s}{s_0} \quad \text{if } s > s_0.$$

We can choose any constant value of k to suit our units. If s is measured in Watts/m^2, then normally k is taken to be 1. The unit of sensation (of sound) measured is then called *Bel* (this unit is named after the great 19th century American inventor Alexander Graham Bell). The tenth part of a Bel is called deciBel (dB). As $s_0 = 10^{-12}$, therefore, $t = \log s - \log s_0 = 12 + \log s$ Bels, or

$$t = 120 + 10 \log s \text{ dB}.$$

The relation between a stimulus and the corresponding sensation (in the case of sound) will be seen in the table [shown on the following page].

Note that as the stimulus increases from 10^{-11} to 1 (a hundred billion-fold increase), the sensation increases only 12 fold (from 1 to 12 B or 10 to 120 dB). Our eardrums are so delicate that even a 12-fold increase is just bearable; if we had to endure even 20 times as much sensation as a barely audible sound, our eardrums would burst and we would become deaf. So how nice of nature to "logarithmize" all stimuli before "sensationalizing" them!

Sound Stimulus (Watts/m^2)	Sound Sensation (dB)
10^{-12}	0
10^{-11}	10
10^{-6}	60
1	120

The units of measurement, which are based on logarithms of normal units, are called logarithmic scales of measurement. Bel (or deciBel) is a logarithmic scale for measuring sound sensation.

You will now be interested in knowing the dB measure of some sounds.

Sound	Sensation (dB)
Humming (to oneself)	10
Normal conversation	60
Thunder	120

Different makes of aircraft create different intensities of sound when taking off, cruising or landing. Their dB measures are of the following order.

Aircraft type	Taking off (dB)	Cruising (dB)	Landing (dB)
1. Avro	92.5	96.4	103.8
2. Fokker Friendship	92.4	98.6	96.6
3. Airbus A-300	95.3	101.6	100.0
4. Boeing 707	114.4	108.0	120.0
5. Boeing 737	96.0	104.0	108.0
6. Boeing 747 (Jumbo)	107.0	100.0	107.0
7. Concorde	117.8	113.1	114.9

Let us work out a sum on the basis of this information. The Concorde is a supersonic aircraft, i.e., it flies faster than sound. Let us find the ratio of sound (stimulus) produced by a Concorde and a Jumbo jet when taking off. Suppose the intensities of sounds produced by a Concorde and Jumbo are s_1 and s_2, respectively.

Then in dB,

$$117.8 = 120 + 10 \log s_1,$$
$$107.0 = 120 + 10 \log s_2.$$

So $\log s_1 - \log s_2 = 1.08$ or $\log(s_1/s_2) = 1.08$. That is,

$$s_1/s_2 = 10^{1.08} = 12$$

approximately. Thus, Concorde produces 12 times more sound than a Jumbo jet. Because of the "logarithmisation", our ears are able to bear the big sound produced by a Concorde, but what about inanimate objects like glass window-panes of the airport building and of buildings near the airport? This is why Concordes are not welcome at all airports. In fact, there are very few airports in the world where Concordes are permitted to land.

Acknowledgement The material in Section 8.5 has been reproduced from *Bona Mathematica*, Volume 6, Numbers 1–3, by kind permission from the publisher (Bhaskaracharya Pratishtana, 56/14, Erandvane, Damle Path, Off Law College Road, Pune–411 004, Maharashtra, India).

Chapter 9

Three Applications

9.1 Starry starry night

> *Twinkle twinkle little star,*
> *How I wonder what you are.*
> *Up above the world so high,*
> *Like a diamond in the sky.*

So runs a popular children's ditty. In this section, we shall see how astronomers express the brightness of a star as a number, and how logarithms play a role in this description.

The brightest and perhaps most beautiful of stars is Sirius. It is brilliantly visible in the winter sky, shining like a diamond in the constellation of Canis Major, which lies adjacent to the most recognizable of all constellations—Orion the Hunter. It is also known as the Dog Star (the dog, naturally, belongs to the Hunter). The next brightest, and as beautiful, is Canopus, which dominates the southern sky, lying close to the Centaurus group of stars—α Centauri, β Centauri, . . .—these being the stars closest to the Earth).

The system of classification of stars according to brightness goes back to the ancient Greek astronomer Hipparchus, who classified the visible stars into six classes. Amazingly, his system of classification survives to this day. What he did was to list the twenty brightest stars as stars of the *first magnitude* (Sirius and Canopus belong to this list); then he listed the next fifty brightest stars as stars of the *second magnitude*; and so on, down to stars of the *sixth magnitude*, which comprised several hundred rather faint stars barely visible to the naked eye.

Note that the system is based on *apparent brightness*. Obviously, stars that are far away will appear dim, even though they may be extremely bright. Astronomers refer to the brightness of a star as its *magnitude*. So we need to distinguish between the *apparent magnitude* of a star, which indicates how bright it appears to an observer on the Earth, and the *absolute magnitude*, which indicates how bright it actually is; that is, how bright it would look from some fixed distance (the same for all stars). Thus, the absolute brightness of a star measures its "true brightness".

As time passed, the decimal division of magnitudes came into use. In this system, a star half-way in brightness between a star of magnitude 2 and a star of magnitude 3 is assigned a magnitude of 2.5; a star whose apparent brightness is a little less than that of a star of magnitude 3 is assigned a magnitude of 2.9; and so on. As the system gained in favour, the classification scheme became finer, and stars which had earlier been listed as first magnitude stars were reclassified as having magnitudes of 0.9, 0.8, 0.7, and so on, according to the differences between them.

In the mid-1800s, the system of classification became finer still as instruments became available that could accurately measure the light intensity or *light flux* of a source—the light energy received per unit time from the source. With the help of such instruments a curious discovery was made, namely that:

- the light flux of a star of magnitude 2 is 2.512 times as much as that of a star of magnitude 3;

- the light flux of a star of magnitude 2 is 2.512^2, or 6.31 times as much as that of a star of magnitude 4, and 2.512^3, or 15.85 times as much as the light flux of a star of magnitude 5;

- the light flux of a first magnitude star is 100 times the light flux of a sixth magnitude star (observe that 2.512^5 is very close to 100); and so on.

This is clearly related to Fechner's law: *"The response of the senses varies as the logarithm of the stimulus."* From this, we deduce the following:

> If the difference in apparent magnitudes between two stars is *n*, then the ratio of their light fluxes is 2.512^n.

Otherwise expressed, if the magnitudes of two stars A and B are m_A and m_B, respectively, and their light fluxes are l_A and l_B, respectively, then

$$m_A - m_B = 2.512 \log_{10} \frac{l_B}{l_A}.$$

Observe how logarithms enter into the picture quite naturally at this point.

The *absolute magnitude* of a star is a measure of how bright the star would appear if it were at a distance of 10 parsecs from the Earth. (One parsec is roughly 3.26 light years. The word 'parsec' comes from a combination of 'parallax' and 'second of arc'.) Clearly, to measure the absolute magnitude of a star, we need to know both its apparent magnitude and its distance from the Earth. This means that we must have at hand techniques for measuring the distances of interstellar objects. Such techniques are available and have been known to astronomers from ancient times.

Table 9.1 gives the apparent magnitude (m) of some familiar stellar objects. Observe that the more negative the magnitude, the brighter the object. The Sun, with a magnitude of -26.5, is the brightest object visible to us; after that comes the full Moon (magnitude -12.5). The difference of 14 in the magnitudes of the Sun and Moon means that the light flux from the Sun is 2.512^{14} or approximately 400000 times the light flux from the Moon. Polaris, also known as the North Star, has a magnitude of roughly 2.1, which means that the light flux from the Moon is $2.512^{14.6}$ or roughly 700000 times the light flux from Polaris.

The Absolute Magnitude of the Sun may be computed from this data as follows. The Earth–Sun distance is, by definition, equal to

Table 9.1 *Apparent brightness of some familiar objects*

Object	m
Sun	-26.5
Full Moon	-12.5
Venus (maximum)	-4.0
Sirius	-1.5
Naked-eye limit	$+6.5$
Binocular limit	$+10.0$
15-cm telescope limit	$+13.0$
Photographic limit	$+24.0$

Table 9.2 *Magnitudes of some familiar stars*

Object	m	M
Sirius	−1.46	+1.4
Canopus	−0.72	−3.1
α Centauri	−0.01	+4.4
Arcturus	−0.06	+0.5
Vega	+0.04	+0.5
Capella	+0.05	−0.7
Rigel	+0.14	−6.8
Procyon	+0.37	+2.6
Betelguese	+0.41	−5.5
Aldebaran	+0.86	−0.2
Antares	+0.92	−4.5
Pollux	+1.16	+0.8
Deneb	+1.26	−6.9

1 AU, with 1 AU being approximately 150×10^6 kilometers. Now, 1 parsec is roughly 2×10^5 AU, so if the Sun were at a distance of 10 parsecs or 2×10^6 AU, its brightness would decrease by a factor of $(2 \times 10^6)^2$ or 4×10^{12}. We must now find n such that

$$2.512^n \approx 4 \times 10^{12}.$$

Taking logarithms to base 10, we obtain

$$n \approx \frac{\log_{10} 4 \times 10^{12}}{\log_{10} 2.512} = \frac{12.6021}{0.4000} \approx 31.5.$$

Since the apparent magnitude of the Sun is −26.5, we see that its absolute magnitude M is $-26.5 + 31.5 = +5.0$; so the Sun is a mere fifth magnitude star! (More accurate computations give the figure +4.8 for the magnitude of the Sun.)

Table 9.2 gives the apparent magnitudes (m) and absolute magnitudes (M) of a few familiar stars.

Note that Rigel (in Orion) and Deneb are, in absolute terms, the brightest objects in this list. Proxima Centauri, the closest star (its distance from the Earth is 1.31 parsecs), has $m = +11.05$ and $M = +15.4$, which make it a rather insignificant star, distinguished only by its closeness to us. The next closest star, α Centauri, at a distance of 1.35 parsecs, has $m = -0.01$ and $M = +4.4$.

EXERCISE

9.1.1 Let m and M be the apparent and absolute magnitudes of a star whose distance from the Earth is r (measured in light years). Show that we have the approximate relation

$$m - M = 5\log_{10}\frac{r}{10}.$$

9.2 Earthquake!

An earthquake happens when rocks within the Earth's crust either fracture or slip against one another. The shock waves produced by the event travel through the Earth at high speed and can cause immense damage to houses, buildings, bridges and other such man-made structures; the death toll can be very high. The point on the Earth's surface directly above the point of fracture or slippage is called the *epicenter* of the earthquake, and it is here that maximum damage occurs. The larger the slippage, the greater the energy released during the earthquake and the greater the damage. The earthquake itself lasts for a few minutes at most (usually for just a few seconds), but the waves generated during this short span of time are enough to destroy the foundations of buildings and bridges.

When the epicenter of the earthquake lies underneath an ocean, say the Atlantic Ocean (near or on the mid-Atlantic Ridge) or the Pacific Ocean, the effect frequently is a *tidal wave*. This is the wave created by the slippage of the sea floor. Incredibly high tidal waves have been reported in countries such as Japan—up to 60 feet or more. The destruction caused when such waves reach the shore and hit human settlements can only be imagined.

★ ★ ★

COMMENT. Not all earthquakes arise from slippage of rock strata against one another; some arise because of the curious phenomenon of expansion or contraction that accompanies a change of phase. Just as contraction accompanies the melting of ice (and, conversely, expansion accompanies the formation of ice), so also such changes of volume accompany the phase changes that occur within the molten magma within the Earth. These, in turn, may be caused by changes of pressure. Whatever be the cause, the effect of a sudden change of volume is to cause a shock wave to be transmitted outwards, and an earthquake results.

9.2.1 Seismology

The waves radiating out from the epicenter are known as *seismic waves*, and they can be measured very accurately using *seismometers*. Geologists routinely place seismometers at strategic spots on the Earth's surface. When an earthquake takes place, the instruments measure the size of the waves passing through the Earth's surface at that instant. By comparing the measurements from seismometers located at different spots, the epicenter of the earthquake can be worked out very precisely, as also the strength of the earthquake. A study of the wave patterns also reveals a lot about the internal structure of the Earth: for instance, the densities and thicknesses of the various layers under the crust. Using such techniques, we can obtain information about the solid core of the earth and the molten portions, and the boundary separating the two.

★ ★ ★

COMMENT. Seismometer technology has at least one unexpected but very important consequence: it allows countries to find out details about the nuclear-weapons testing program of other countries. This is because when a nuclear weapon is detonated, even a test weapon, it sends out waves just like an actual earthquake. The analysis of the measurements helps in finding the strength of the weapon as well as the spot where the test has been carried out.

Another curious use to which seismometer technology has been put is in understanding the internal structure of the Moon, by studying the vibrations created within it when an object collides with it. Since astronomers cannot arrange for collisions between the moon and heavy meteors (and thank God for that!), they have done the next best thing: that of sending discarded space vessels (from space missions) crashing down onto the surface of the Moon. Seismometers left behind on the Moon by earlier space missions now record the relevant data, which are then sent to the control mission on the Earth for analysis.

9.2.2 The Richter scale

Various numerical measures have been devised for categorizing the intensity of an earthquake. The best known amongst these is the RICHTER SCALE, which uses the size of the seismic waves generated by the earthquake. A quake measuring 7.0 or higher on the Richter scale would have very serious consequences. Another scale often used is the MODIFIED MERCALLI SCALE, which ranks an earthquake according to the effect on human settlements (intensity I: can barely be felt; intensity XII: total damage).

The Richter scale is logarithmic: if r is the Richter magnitude of an earthquake in which the energy released is E, then $r = \log_{10} E$ plus some appropriate constant. So an earthquake with $r = 8.0$ releases 10 times as much energy as an earthquake with $r = 7.0$, and 100 times as much energy as an earthquake with $r = 6.0$. An earthquake with r in the range 2–4 is barely felt, while one with r in the range 4–5 is quite perceptible. If r is in the range 5–6, some damage to buildings will occur, and if r is in the range 6–7, there will be considerable damage to buildings. An earthquake with r above 8 will cause extremely serious damage.

Listed in Table 9.3 are some of the worst earthquakes of the 1900s (r refers to the Richter magnitude of the quake). The last one listed was one of the most destructive in recorded history, with a death toll of nearly 750,000; but there is one which occurred in 1556 in Shenshi, China whose death toll at 830,000 was still higher (Richter magnitude unknown).

★ ★ ★

It has been estimated that about 800000 earthquakes occur each year with r in the range 2.0–3.4; about 35000 per year with r in

Table 9.3 *Magnitudes of some of the worst earthquakes*

Date	Location	Death Toll	r
Apr 1905	Himachal Pradesh, India	375,000	8.3
Apr 1906	San Francisco (California)	700	8.3
Dec 1920	Kansu, China	100,000	8.6
Sep 1923	Sagami Bay, Japan	100,000	8.3
Aug 1950	Assam, India	30,000	8.4
May 1960	Arauco, Chile	2,000	8.4
Mar 1964	Anchorage, Alaska	114	8.5
Jul 1976	Northeastern China	695,000	7.9

Table 9.4   *The Beaufort scale for wind speed*

Beaufort number	Label	Wind speed
0	Calm	0–1 knots
1	Light air	1–3 knots
2	Light breeze	4–6 knots
4	Moderate breeze	11–16 knots
5	Stiff breeze	17–21 knots
6	Strong breeze	22–27 knots
7	Moderate gale	28–33 knots
8	Fresh gale	34–40 knots
9	Strong gale	41–47 knots
10	Whole gale	48–55 knots
11	Storm	56–63 knots
12	Hurricane	64-plus knots

the range 3.5–4.8; about 2000 per year with r in the range 4.9–6.1; about 100 per year with r in the range 6.2–6.9; about 20 per year with r in the range 7–8; and about 10 or so per century with r above 8.

9.2.3   Storm warning

The Richter and Mercalli scales for assessing the power of an earthquake may remind readers about the system (due to Admiral Beaufort) used to describe wind speed. Table 9.4 gives the details. The scale used is the *Beaufort Scale*. Here is how it is read. Speeds are in *knots*, one knot being a speed of 1 nautical mile per hour (1.15 miles per hour, or 1.85 km per hour).

For completeness, we list in Table 9.5 some visual clues that help in estimating wind speed. Note that this is a purely visual system, with no precise quantitative basis.

9.3   Sweet and sour

Chemists frequently need to describe how acidic or alkaline a particular solution is; i.e., to give a quantitative measure of its concentration of hydrogen ions, H^+. The scale generally used is the pH scale; pH means *potential for hydrogen*.

Table 9.5 *Visual clues to help in assessing wind speed*

Wind speed	Description
0–1 knots	Smoke rises vertically; sea is mirror-smooth
1–3 knots	Smoke shows direction of wind
4–6 knots	Wind felt on face; leaves rustle in trees
11–16 knots	Loose paper blows around; frequent whitecaps on the sea
22–27 knots	Whistling of telephone wires; spray felt on the sea surface
28–33 knots	Large trees sway
34–40 knots	Twigs break from trees; streaks of foam on the sea
41–47 knots	Branches break from trees
48–55 knots	Trees are uprooted; sea takes on white appearance
56–63 knots	Widespread damage
64-plus knots	Storm waves at sea

The pH ranges from 0 to 14; a pH below 7 indicates that the solution is acidic, and a pH greater than 7 indicates that the solution is alkaline. A neutral solution such as pure water has a pH of 7. (Strictly speaking, the temperature must be mentioned when we quote the pH value. For pure water, pH equals 7 only at 25°C.) Full strength sulphuric acid (H_2SO_4) has a pH of 0, as does concentrated nitric acid (HNO_3), or concentrated hydrochloric acid (HCl).

The pH scale was invented in 1909, by a Dutch biochemist, Søren Sørensen. The scale is logarithmic in nature; pH is defined by the equation

$$pH = -\log_{10}\left[H^+\right],$$

where H^+ refers to the concentration of H^+ ions, expressed in moles per liter. A solution with a pH of 4 contains 10^{-4} moles of H^+ ions per liter, and so on.

The pH may be measured by using an electronic pH meter, or by means of special dyes called acid-base indicators. The dyes respond to the acidity level by changing colour in a way that depends on the pH. Another measuring device is pH paper. This has indicators within it, which change colour at different pHs. When dipped into a solution, the paper colour indicates the approximate pH of the solution.

Many chemical reactions depend on pH; indeed, they may be highly sensitive to pH. Agricultural practices have to take soil pH into consideration, because certain crops grow best in an acidic soil (pH < 7), whereas others grow best in an alkaline soil (pH > 7). When the soil is found to be too acidic, lime may be added to make it more alkaline ("sweeten it", to use the gardener's terminology). The fallen leaves of certain trees contribute to the soil's acidity and the consequent lack of growth of plants in the shade of the tree. This happens, for example, with eucalyptus trees.

Chapter 10

The Number e

10.1 Genesis of e

We have encountered the number e earlier in the book. In this chapter and the next one, we make a rather more formal study of this extraordinary number, which has a way of cropping up everywhere, in situations that seem to have little or nothing to do with logarithms.

The simplest way of defining e is via the sequence of numbers

$$(1+1)^1, \quad \left(1+\frac{1}{2}\right)^2, \quad \left(1+\frac{1}{3}\right)^3, \quad \left(1+\frac{1}{4}\right)^4, \quad \ldots.$$

These numbers arise when we compute the growth of a sum of money which is earning interest in a bank, with the compounding being done several times a year at equal intervals of time. If the sum deposited initially is P, the annual interest rate is i, and compounding is done n times a year (every $1/n$ of a year), then the sum of money at the end of the year is

$$P\left(1+\frac{r}{n}\right)^n,$$

where $r = i/100$. If we take a value of 1 for P and a value of 100% for i (though that may be a bit optimistic!), then $r = 1$, and the sum of money at the end of 1 year is given by a_n where $a_n = (1+1/n)^n$.

It is not hard to anticipate that a_n increases with increasing n; for with compounding done more frequently, clearly we must

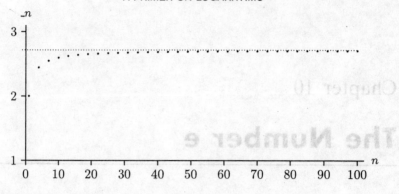

Figure 10.1. *Graph of a_n*

benefit. The figures certainly bear this out:

$$a_2 = 2.2500, \qquad a_3 \approx 2.3704,$$
$$a_4 \approx 2.4414, \qquad a_5 \approx 2.4883,$$
$$a_{10} \approx 2.5937, \qquad a_{20} \approx 2.6533.$$

Note the gradual increase in the values of a_n.

It may seem now that raising the value of n indefinitely will result in indefinitely large gains to us; but, surprisingly, this is not so. Once again, here are a few figures:

$$a_{100} \approx 2.70481, \qquad a_{500} \approx 2.71557,$$
$$a_{1000} \approx 2.71692, \qquad a_{5000} \approx 2.71801.$$

There seems to be a barrier which we are unable to penetrate!

A graphical plot of a_n against n will help us to see the nature of the barrier a little more clearly. Figure 10.1 has been drawn with this purpose in mind.

Note the very visible 'leveling off'. The dotted line suggests a 'barrier' of about 2.718 or so. *It is precisely this number which is denoted by e.*

More sophisticated calculations show that the value of e to 14 decimal places is given by

$$e = 2.71828182845904.$$

10.2 A graphical approach

There is another way in which e may be defined; it involves the use of graphs. We consider the graphs of a^x for various values of

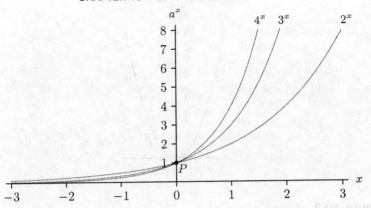

Figure 10.2. *Graph of a^x for $a = 2, 3$ and 4*

a: 2, 2.5, 3, 3.5, and so on. The range of x we are be interested in is -3 to 3 or so. It is not hard to see that as a increases,

- to the right of $x = 0$ the curves get steadily steeper; and

- to the left of $x = 0$ the curves drop down to the x-axis much faster.

The graphs shown in Figure 10.2 correspond to $a = 2$, $a = 3$ and $a = 4$.

Observe that the curves pass through the point $P(0, 1)$ (naturally, $a^0 = 1$ for all $a > 0$); this is the 'cross-over point'. Clearly, the curves corresponding to larger values of a are steeper at P than those with smaller values of a.

Remarks on the concept of a slope   Before proceeding, it may be worthwhile to make some remarks about the concept of a 'slope'. The *slope* of a curve at a point P is a measure of how steep it is at P; that is, with reference to the coordinate system in which the curve is placed, how fast the y-coordinate of a point travelling along the curve in the immediate vicinity of P is increasing with respect to the x-coordinate. One way to estimate the slope at P is to consider a tiny chord PQ of the curve; one end-point of the chord is P and the other end is a point Q, also on the curve, very close to P. Next, we find the slope of the chord; that is, the ratio

$$\frac{y_Q - y_P}{x_Q - x_P},$$

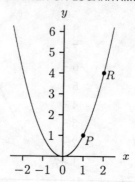

Figure 10.3. *Graph of* $y = x^2$

where $P = (x_P, y_P)$ and $Q = (x_Q, y_Q)$. If the difference in the x-coordinates of P and Q is small enough, i.e., $x_P - x_Q$ is close to 0, then this ratio will give us a good idea of the slope of the curve at P.

Sample computation of slope: for a parabola Consider the curve with equation $y = x^2$. This curve is a *parabola*; it appears as shown in Figure 10.3. We focus attention on the point $P(1,1)$ on the curve. How do we find the slope of the curve at P?

Following the comments made above, we select points Q on the curve that are close to P, and then we find the slopes of the chords PQ. Here are the results:

- if $Q = (1.1, 1.1^2)$ or $(1.1, 1.21)$, then the slope of PQ is $(1.21 - 1.1)/(1.1 - 1)$ or 2.1;

- if $Q = (1.01, 1.01^2)$ or $(1.01, 1.0201)$, then the slope of PQ is 2.01;

- if $Q = (1.001, 1.001^2)$ or $(1.001, 1.002001)$, then the slope of PQ is 2.001.

It is not hard to spot the pattern: if we take the point Q to be $(1 + h, (1 + h)^2)$ where h is small (close to 0), then the slope of PQ is $2 + h$. This quantity can be made indefinitely close to 2, simply by taking h to be small enough, so we take the obvious step and declare the slope of the curve at P to be equal to 2.

We may similarly estimate the slope of the curve at the point $R = (2, 4)$. We select points S on the curve, close to R, and compute the slopes of the chords RS.

Here is what we get:

- if $S = (2.1, 2.1^2)$ or $(2.1, 4.41)$, then the slope of RS is 4.1;

- if $S = (2.01, 2.01^2)$ or $(2.01, 4.0401)$, then the slope of RS is 4.01;

- if $S = (2.1, 2.001^2)$ or $(2.001, 4.004001)$, then the slope of RS is 4.001.

As earlier, the pattern is clear: if $S = (2 + h, (2 + h)^2)$ where h is small, then the slope of RS is $4 + h$. This quantity can be made indefinitely close to 4, by taking h to be small enough, so we declare the slope of the curve at R to be equal to 4.

Computation of slopes for exponential curves We shall use the same strategy for the $y = a^x$ curves. Thus, to estimate the slope of the $y = 2^x$ curve at the point $P = (0, 1)$, we select several nearby points Q on the curve, e.g., $Q = (0.01, 2^{0.01})$, $Q = (0.001, 2^{0.001})$, and so on. Next we compute the slopes of these chords; that is, we compute values such as

$$\frac{2^{0.01} - 1}{0.01}, \quad \frac{2^{0.001} - 1}{0.001}, \quad \frac{2^{0.0001} - 1}{0.0001}, \quad \frac{2^{0.00001} - 1}{0.00001},$$

and so on. We find the values to be

$$0.69556, \quad 0.69339, \quad 0.69317, \quad 0.69315,$$

respectively. It seems reasonable, therefore, to take the slope at P of the 2^x curve to be about 0.69315.

We may similarly estimate the slopes of the other two curves (at the point P). For the curve $y = 3^x$ we take the slope to be approximately

$$\frac{3^{0.00001} - 1}{0.00001} \approx 1.09862,$$

and for the curve $y = 4^x$ we take the slope to be approximately

$$\frac{4^{0.00001} - 1}{0.00001} \approx 1.3863.$$

Let us write m_a to denote the slope of the curve $y = a^x$ at the point $P = (0, 1)$ (in coordinate geometry the letter m is frequently used to denote slope). We can estimate the value of

m_a for any value of a, by mimicking what was done above. For instance, for $a = 2.5$ we get $m_a \approx 0.916295$, this being the value of $(2.5^{0.00001} - 1)/0.00001$.

The values of m_a obtained till now may be displayed in the form of a table as shown below.

a	2	2.5	3	4
m_a	0.69315	0.91629	1.09862	1.3863

Observe that as a increases, so does m_a. Since $m_a < 1$ for $a = 2$ and $a = 2.5$, and $m_a > 1$ for $a = 3$ and $a = 4$, it seems reasonable to suppose that m_a will take the value 1 for *some* value of a between 2.5 and 3. It can be shown (by careful arguments, which we shall not go into here) that this is so. Moreover, we may expect the value of a for which $m_a = 1$ to be roughly half-way between 2.5 and 3. In fact, we find the following:

The value of a for which $m_a = 1$ is e.

Does this come as a major surprise? Perhaps it does, at any rate to the young student who has not encountered e before. We shall now spoil the surprise somewhat by showing that actually there is not very much cause for surprise.

10.3 Explanations

Recall that we had earlier defined e to be the "barrier" to the sequence

$$(1 + 1)^1, \quad \left(1 + \frac{1}{2}\right)^2, \quad \left(1 + \frac{1}{3}\right)^3, \quad \left(1 + \frac{1}{4}\right)^4, \ldots$$

Writing a_n for $(1 + 1/n)^n$ we had found that

$$a_{1000} \approx 2.71692, \quad a_{5000} \approx 2.71801,$$

and that with further increases in n, the value of a_n does not change very much. So e is roughly 2.718.

The very important inference we draw from this is the following:

If n is indefinitely large, then $\left(1 + \dfrac{1}{n}\right)^n \approx e.$

The phrase "indefinitely large" may seem ambiguous (who, after all, is to judge how large is 'indefinitely large'?), but what we mean is that as n gets larger and larger, the approximation to e given by $(1 + 1/n)^n$ gets better and better, in the sense that the ratio $(1 + 1/n)^n : e$ gets closer and closer to $1 : 1$; and "in the limit" we have equality.

Taking the nth roots on both sides of the above relation, we deduce the following:

If n is indefinitely large, then $\quad e^{1/n} \approx 1 + \dfrac{1}{n}.$

Now when n is very large, $1/n$ is very small. Write h for $1/n$; then h is close to 0, and we may rewrite the above as follows:

If h is close to 0, then $\quad e^h \approx 1 + h.$

As earlier, you may feel uncomfortable with the ambiguity in the phrase "close to 0", but what we mean, once again, is that the approximate relation gets better and better as h gets closer and closer to 0 (so the ratio $e^h : 1 + h$ gets closer and closer to $1 : 1$); in the limit, of course, we have equality.

Recall that in the graphical approach, above, we were interested in the slope of the curve $y = a^x$ at the point $P = (0, 1)$. In particular, we wanted the value of a for which this slope is 1. We shall now show that when $a = e$, the slope is 1 (i.e., $m_e = 1$).

Let Q be another point on the curve $y = e^x$, close to P; say $Q = (h, e^h)$, where h is small. The slope of the chord PQ is m_{PQ} where

$$m_{PQ} = \frac{e^h - 1}{h - 0}.$$

Since $e^h \approx 1 + h$, the above fraction approximates to

$$\frac{(1 + h) - 1}{h} = 1.$$

So when h is small, the slope of PQ is close to 1; and it may be made as close to 1 as we want by taking h to be small enough. We conclude the following:

The slope of the $y = e^x$ curve at the point $(0, 1)$ is 1.

This is what we had stated above. So the mysterious coincidence is not really a coincidence!

10.4 Natural logarithms

Having introduced e we now explain why the logarithm to base e possesses a certain "naturalness", much more so than the 'common logarithm', whose base is 10.

Notation Logarithms to base e are denoted by the word 'ln' and are known as 'natural logarithms'; so $\ln x$ carries the same meaning as $\log_e x$, and we have

$$y = \ln x \iff x = e^y.$$

The notation reflects the importance of this function: it occurs sufficiently often in engineering and the physical sciences that a special notation was devised for it.

Common logarithms were devised in the 1600s primarily as an aid to computation, but this need is now playing itself out, thanks to the modern electronic calculator. However, this cannot happen to the natural logarithm function—it possesses a fundamental significance, independent of computational or other needs; so it will *always* be of relevance. Some reasons for this will appear in Chapters 11 and 12.

A few values of the ln function are listed below:

$$\ln 2 = 0.69315, \quad \ln 3 = 1.09861,$$
$$\ln 5 = 1.60944, \quad \ln 10 = 2.30259.$$

The approximation $\ln 10 \approx 2.30259$ means that $\log_{10} e$ is approximately equal to $1/2.30259$, i.e., $\log_{10} e \approx 0.4343$. Since $\log_{10} x = \ln x / \ln 10$, we get

$$\log_{10} x \approx 0.4343 \ln x.$$

This relation was mentioned in Section 5.5.

$$\star \star \star$$

Various results hold for the natural logarithm function. We list a few of these results below.

(1) If h is very close to 0, then $\ln(1 + h) \approx h$.

> **EXAMPLE.** We have, $\ln 1.01 \approx 0.00995$, which is nearly equal to 0.01, and $\ln 1.002 \approx 0.001998$, which is nearly equal to 0.002.
>
> It may be shown that a superior approximation is given by $\ln(1 + h) \approx h - h^2/2$:

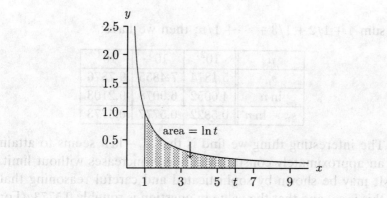

Figure 10.4. *Graph of $y = 1/x$*

(2) Let a be any positive real number; then the slope of the curve $y = a^x$ at the point $P(0, 1)$ is equal to $\ln a$. That is, the number m_a defined earlier (Section 10.2) is, in fact, equal to $\ln a$.

(3) The area enclosed by the curve $y = 1/x$, the x-axis and the lines $x = 1$ and $x = t$ is equal to $\ln t$.

The curve and region in question are sketched in Figure 10.4.

(4) Let $n > 0$ be a large integer; then we have the following double inequality (the sums on the left and right both have $n - 1$ terms):

$$\frac{1}{2} + \frac{1}{3} + \cdots + \frac{1}{n} < \ln n < 1 + \frac{1}{2} + \cdots + \frac{1}{n-1}.$$

From this, we quickly deduce that

$$\ln n + \frac{1}{n} < 1 + \frac{1}{2} + \frac{1}{3} + \cdots + \frac{1}{n} < \ln n + 1.$$

So the sum $1 + 1/2 + 1/3 + \cdots + 1/n$ lies between $\ln n + 1/n$ and $\ln n + 1$. We may now make a rough guess that

$$1 + \frac{1}{2} + \frac{1}{3} + \cdots + \frac{1}{n} \approx \ln n + 0.5.$$

This result turns out to be fairly close to the truth. For comparison, we offer the following data. Let s_n denote the sum $1 + 1/2 + 1/3 + \cdots + 1/n$; then we have:

n	10^2	10^3	10^4
s_n	5.1874	7.4855	9.7876
$\ln n$	4.6052	6.9078	9.2103
$s_n - \ln n$	0.5822	0.5777	0.5773

The interesting thing we find is that $s_n - \ln n$ seems to attain an approximately constant value as n increases without limit. It may be shown by sophisticated and careful reasoning that this is so, and that the value in question is roughly 0.5773. (For comparison, when $n = 10^5$ the difference between s_n and $\ln n$ is 0.57723.) This constant is known as the *Euler–Mascheroni constant*. So we have the following nice result:

When n is large, $s_n \approx \ln n + 0.5773$.

COMMENT. The reader may wonder whether there is any 'exact' formula for s_n. Why should we seek approximate formulas when there may be an exact one? It turns out that no such exact formula exists; at any rate, no formula expressed in terms of quantities that are familiar and easy to compute. There are many instances where this kind of thing happens, i.e., where we have no worthwhile and exact formula for a sum. In such cases, an easy-to-use, approximate formula is much more valuable than an exact formula which is contrived and hard-to-use.

(5) Let $n > 0$ be a large integer; then the sum

$$\frac{1}{n+1} + \frac{1}{n+2} + \frac{1}{n+3} + \cdots + \frac{1}{2n}$$

is approximately equal to $\ln 2$; and the approximation improves as n increases.

EXAMPLE. For $n = 1000$ the sum is roughly 0.6929, which is close to 0.69315 (the actual value of $\ln 2$).

(6) Let $n > 0$ be a large integer; then the sum

$$\frac{1}{n+1} + \frac{1}{n+2} + \frac{1}{n+3} + \cdots + \frac{1}{3n}$$

is approximately equal to $\ln 3$; and the approximation improves as n increases.

EXAMPLE. For $n = 1000$ the sum is roughly 1.0983, which is close to 1.09861 (the actual value of $\ln 3$).

(7) More generally, let k be any positive integer greater than 1; and let $n > 0$ be a large integer, then the sum

$$\frac{1}{n+1} + \frac{1}{n+2} + \frac{1}{n+3} + \cdots + \frac{1}{kn}$$

is approximately equal to $\ln k$; and the approximation improves as n increases.

(8) Let $n > 0$ be a large integer; then the sum

$$1 - \frac{1}{2} + \frac{1}{3} - \frac{1}{4} + \cdots + \frac{1}{2n-1} - \frac{1}{2n}$$

is approximately equal to $\ln 2$; and the approximation improves as n increases.

EXAMPLE. For $n = 100$ the sum is 0.69065, and for $n = 1000$ the sum is 0.6929; contrast these values with the true value of $\ln 2$ (roughly 0.69315).

(9) Let $n > 0$ be large; then $\ln(n + 1) - \ln n$ is approximately equal to $1/n$.

This holds because $\ln(n + 1) - \ln n = \ln \dfrac{n+1}{n}$, and

$$\ln \frac{n+1}{n} = \ln \left(1 + \frac{1}{n}\right) \approx \frac{1}{n},$$

since n is large (implying that $1/n$ is small; now use result (1)).

This result is of vital significance, because it shows that the rate of growth of $\ln x$ is inversely proportional to x. There are many natural phenomena for which this property holds, and the logarithmic function enters in an inevitable manner into the analysis of such phenomena.

The reader may note the close interconnection between the results (3)–(7) and (9).

EXERCISES

10.4.1 Examine computationally the sequence whose n^{th} term is given by

$$\left(1 + \frac{1}{2n}\right)^n.$$

That is, compute the values of the quantities

$$\left(1 + \frac{1}{2}\right)^1, \quad \left(1 + \frac{1}{4}\right)^2, \quad \left(1 + \frac{1}{6}\right)^3, \quad \left(1 + \frac{1}{8}\right)^4, \quad \dots.$$

You will find that the quantities seem to approach some limiting value. How is this number related to e?

10.4.2 Do the same with the sequence whose n^{th} term is given by

$$\left(1 + \frac{1}{3n}\right)^n.$$

10.4.3 Do the same with the sequence whose n^{th} term is given by

$$\left(1 + \frac{2}{n}\right)^n.$$

10.4.4 Find a generalization based on the results of the preceding three problems.

Chapter 11

The Mystique of e

Let's face it: the number e *has* a mystique. It has a habit of cropping up all over the place, in very unexpected ways, much like its illustrious (and more well-known) cousin, π. Though π and e have rather different origins, they are related to one another in many unusual ways; they are indeed first cousins! We shall list some of these connections in this chapter. Some of these 'connections' involve concepts that have not been developed in this book (e.g., probability, expectation), so references to them will be rather brief. The reader is invited to explore some of these connections, using source material from the reference list.

Before proceeding, we remark that the sequence a_1, a_2, a_3, a_4, ... given by

$$a_n = \left(1 + \frac{1}{n}\right)^n$$

is a *monotonic increasing sequence*: that is, we have $a_1 < a_2 < a_3 < \cdots$, and the values level off at e; so $a_n < e$ for all n, but the difference between e and a_n shrinks to 0 as n grows without limit.

Curiously, the sequence $b_1, b_2, b_3, \ldots$ defined by

$$b_n = \left(1 + \frac{1}{n}\right)^{n+1},$$

exhibits the opposite behaviour: $\{b_n\}$ is a *monotonic decreasing sequence*, that is, $b_1 > b_2 > b_3 > \cdots$, and $b_n > e$ for all n. As with $\{a_n\}$, however, the difference between e and b_n shrinks to 0 as n

grows without limit. Thus, we have the following figures:

$$b_2 = 3.375, \qquad b_3 \approx 3.1605, \qquad b_4 \approx 3.0518,$$
$$b_{100} \approx 2.73186, \quad b_{200} \approx 2.72507, \quad b_{1000} \approx 2.71964.$$

Observe that the values level off at e, as earlier, but from the opposite direction.

This being the case, we may expect that the AM (arithmetic mean) of a_n and b_n will provide a better estimate of e than either a_n or b_n; and this turns out to be the case. Thus, writing $c_n = (a_n + b_n)/2$, we have

$$c_{10} \approx 2.72343, \quad c_{100} \approx 2.71834, \quad c_{1000} \approx 2.71828.$$

We see that the values get closer and closer to e at a good pace. In fact, we find the following curious pattern:

$$e - c_{100} \approx 5.61 \times 10^{-5},$$
$$e - c_{1000} \approx 5.66 \times 10^{-7},$$
$$e - c_{10000} \approx 5.66 \times 10^{-9}, \quad \ldots.$$

Observe that $c_n > e$ for all the values of n shown. This inequality may be shown to hold for all n.

★ ★ ★

A final remark—e *is an irrational number*. That is, it is not possible to find non-zero integers a and b such that $ae + b = 0$ (nor is it possible to find non-zero integers a, b and c such that $ae^2 + be + c = 0$; see Appendix A).

11.1 Derangements

We perform an experiment with two identical (and complete) packs of cards as follows. We shuffle each pack separately, then place the packs side by side, face down. We now pick up the top card from each pack. The two cards may happen to be different, or they may be the same. If they are the same we record the experiment as a 'success', otherwise we record it as a 'failure'. Then we pick up the next card from each pack, and compare them; once again we record the experiment as a success if they are the same, otherwise we record it as a failure. And so we proceed, till all the cards have been drawn and compared.

What is the chance that on every occasion we obtain a failure? Surprise! The answer turns out to be roughly $1/e$. Moreover, this result holds even if we do the experiment with larger card packs. That is, if we write the numbers $1, 2, 3, \ldots, 99, 100$ on two sets of 100 slips of paper, place the two sets of slips in two vessels, then draw them out at a time from each vessel (each time shuffling the contents thoroughly), then the probability that we obtain 100 failures in a row is roughly $1/e$; and this is so if we replace '100' by '1000'. (Here 'failure' means that the numbers on the two slips drawn are different.)

★ ★ ★

Let us explain what we mean using the language of permutations and derangements.

Consider the word THE. In how many ways can its letters be rearranged? It is easy to check that there are 6 possible ways: THE, TEH, HET, HTE, ETH, EHT. These are the various *permutations* of the letters of the word THE. By experimentation, we find that the letters of a word containing 3 different letters may be permuted in 6 ways. Observe that $6 = 1 \times 2 \times 3$.

Similar experimentation with the word SPIN reveals that its letters may be permuted in 24 ways; and this is so for any word with 4 different letters. (The reader should check this claim.) Observe that $24 = 1 \times 2 \times 3 \times 4$.

In general, the following is true: *n distinct symbols may be arranged in a row ("permuted") in n! different ways, where n!* $= 1 \times 2 \times 3 \times \cdots \times (n-1) \times n$. (Here $n!$ is pronounced as 'n factorial'.)

EXAMPLE. $10! = 1 \times 2 \times 3 \times \cdots \times 9 \times 10 = 3628800$.

Now consider the n numbers $1, 2, 3, \ldots, n-1, n$. We may, as noted, permute them in $n!$ ways. If the permutation is such that every number is in the 'wrong' place, then we refer to it as a *derangement*. We write D_n for the number of derangements. So we have the following:

- Let $n = 3$; then the "word" 123 has the following $3! = 6$ permutations:

$$123, \ 132, \ 213, \ 231, \ 312, \ 321.$$

Of these, 231 and 312 are derangements, and there are no others. So $D_3 = 2$, and the proportion of derangements is $2/6$ in this case.

- Let $n = 4$; the $4! = 24$ permutations of the word 1234 are the following:

$$1234, \quad 1243, \quad 1324, \quad 1342, \quad 1423, \quad 1432,$$
$$2134, \quad 2143, \quad 2314, \quad 2341, \quad 2413, \quad 2431,$$
$$3124, \quad 3142, \quad 3214, \quad 3241, \quad 3412, \quad 3421,$$
$$4123, \quad 4132, \quad 4213, \quad 4231, \quad 4312, \quad 4321.$$

Of these, the derangements are the following: 2143, 2341, 2413, 3142, 3412, 3421, 4123, 4312, 4321. So there are 9 derangements ($D_4 = 9$), and the proportion of derangements is 9/24.

- Let $n = 5$; there are $5! = 120$ permutations of the word 12345. It is tedious to list them, so we invite you to do so! A carefully organized list shows that there are 44 derangements ($D_5 = 44$), and the proportion of derangements is 44/120.

- Let $n = 6$; there are $6! = 720$ permutations of the word 123456. It is considerably more tedious than earlier to list these permutations! It turns out that there are 265 derangements in all, and they are equally distributed with regard to the first digit: 53 derangements have a '2' in the first place, another 53 have a '3' in the first place, and so on; note that $53 \times 5 = 265$. The proportion of derangements is now 265/720.

The reader may wonder whether there is a simple pattern in the sequence $D_1, D_2, D_3, \ldots$ (the sequence of derangement numbers); there is! Examine the data:

n	1	2	3	4	5	6	...
D_n	0	1	2	9	44	265	...

We have placed a 0 and 1 at the start of the sequence, as the number of derangements is 0 when $n = 1$ and 1 when $n = 2$. The pattern is striking:

$$D_2 = 2 \times D_1 + 1$$
$$D_3 = 3 \times D_2 - 1,$$
$$D_4 = 4 \times D_3 + 1,$$
$$D_5 = 5 \times D_4 - 1,$$
$$D_6 = 6 \times D_5 + 1,$$

and so on. The general result clearly seems to be

$$D_n = n\,D_{n-1} + (-1)^n,$$

and it turns out that the relation is true for all $n > 0$. The reader should attempt to prove the result.

Now observe that

$$\frac{2}{6} = 0.333, \quad \frac{9}{24} = 0.375, \quad \frac{44}{120} = 0.367, \quad \frac{265}{720} = 0.368.$$

The three proportions are rather close to one another. Is this a coincidence? Not so! It turns out that for large values of n, the proportion of derangements is nearly equal to $1/e$. (Note that $1/e \approx 0.3679$.) Or, to put another way, the number of derangements is nearly equal to $n!/e$. (In fact, D_n is the integer closest to $n!/e$.) What a surprising and beautiful result!

11.2 The nth root of n

A popular question is: *Which quantity is larger, $\sqrt{2}$ or $\sqrt[3]{3}$?* It is easy to check that the latter quantity is the larger one (to see why, raise both quantities to the sixth power). We find, using similar manipulations, that $3^{1/3}$ exceeds $4^{1/4}$, and $4^{1/4}$ in turn exceeds $5^{1/5}$. The following two questions now pose themselves quite naturally:

- For which positive integer n does $n^{1/n}$ take its largest value?

- For which positive real number x does $x^{1/x}$ take its largest value?

We shall consider the second question. Our analysis will answer the first question at the same time.

Using a computer we compute $x^{1/x}$ for many positive values of x, then we plot a graph of $x^{1/x}$. The result is shown in Figure 11.1.

Observe that the graph is rather flat at the top, for a fairly wide range of x. The evidence is unmistakable: $x^{1/x}$ *seems to take its maximum value at or near e!* We shall now show that this is indeed so. The result proved in the preceding chapter, that

$$\ln(1 + h) \approx h \text{ when } |h| \text{ is very small},$$

will be seen to play a central role in our analysis.

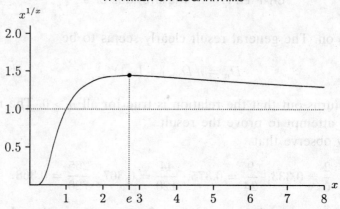

Figure 11.1. *Graph of $x^{1/x}$ for $x > 0$.*

As the expression $x^{1/x}$ is somewhat awkward to handle, we take logarithms to base e and consider instead the expression $f(x)$, where

$$f(x) = \frac{\ln x}{x}.$$

We seek the value of $x > 0$, for which $f(x)$ attains its largest value.

Let x take some value t, and consider what happens when x increases by a small quantity h; i.e., x grows from t to $t + h$; here $h > 0$ but $h \ll t$. (The double inequality sign is a way of saying that t is very small in comparison with t.) Consider the difference produced in the value of f; call it Δf for convenience. We have

$$\Delta f = \frac{\ln(t + h)}{t + h} - \frac{\ln t}{t}.$$

To get a handle on this expression, we first write $\ln(t + h)$ in a more convenient form:

$$\ln(t + h) = \ln\left[t \cdot \left(1 + \frac{h}{t}\right)\right] = \ln t + \ln\left(1 + \frac{h}{t}\right).$$

Since h/t is small, $\ln(1 + h/t)$ may be approximated by h/t; therefore,

$$\Delta f = \frac{\ln(t + h)}{t + h} - \frac{\ln t}{t} = \frac{\ln t + \ln(1 + h/t)}{t + h} - \frac{\ln t}{t}$$

$$\approx \frac{\ln t + h/t}{t + h} - \frac{\ln t}{t} = \frac{t \ln t + h - t \ln t - h \ln t}{t(t + h)}$$

$$= \frac{h(1 - \ln t)}{t(t + h)}.$$

This is a very valuable result! It means that if x changes from t to $t + h$, where h is a tiny positive quantity, then $f(x)$ increases by the amount

$$\frac{h(1 - \ln t)}{t(t + h)}.$$

Observe that the increase is positive if and only if $1 - \ln t > 0$, that is, if and only if $\ln t < 1$. The inequality $\ln t < 1$ means that $t < e$. So we have the following inference:

For $x < e$, increasing x by a small quantity results in an _increase_ in the value of $f(x)$; and for $x > e$, increasing x by a small quantity results in a _decrease_ in the value of $f(x)$.

Observe that our findings are in agreement with the profile suggested by the graph.

The above conclusion implies that the value of $f(x)$ increases steadily till $x = e$, and thereafter it decreases. In other words, $f(x)$ attains its maximum value at $x = e$! So the following statement can now be made with confidence: *For any positive real number x, we have $x^{1/x} \leq e^{1/e}$.*

Note that we have simultaneously established the following:

$$3^{1/3} > 4^{1/4} > 5^{1/5} > 6^{1/6} > 7^{1/7} > 8^{1/8} > \cdots.$$

This holds because $e < 2 < 3 < 4 < 5 < 6 < 7 < 8 < \cdots$. We already know that $3^{1/3} > 2^{1/2}$. Therefore, $n^{1/n}$ attains its largest value when $n = 3$.

The graph in Figure 11.1 suggests something more: that as n increases without limit, $n^{1/n}$ gradually gets closer and closer to 1. The reader should attempt to prove this.

★ ★ ★

Here is another interesting problem, somewhat reminiscent of the problem just considered. We consider the graphs of functions of the variety $f(x) = x\,a^{-x}$ for values of a greater than 1. We sketch the graphs only for non-negative values of x; so the graphs lie in the first quadrant.

Consider, for example, the function $f(x) = x\,2^{-x}$. Clearly, $f(0) = 0$, and $f(x) > 0$ when $x > 0$. So the curve starts out at the origin and climbs into the first quadrant. What happens as x increases without limit? Here we see an interesting tussle: x increases steadily whereas 2^{-x} decreases steadily. Who "wins"? It

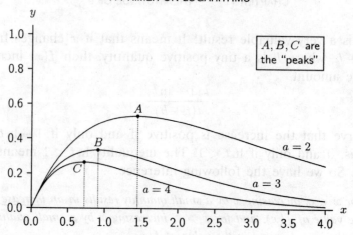

Figure 11.2. *The curve $y = xa^{-x}$ for various a's*

may not be obvious at first, but a bit of experimentation with a calculator quickly confirms that the downward pull of 2^{-x} quickly overwhelms the rising aspirations of x. That is, as x increases, $x\,2^{-x}$ increases for a while, reaches a peak value, then starts to diminish; and thereafter it falls rapidly towards 0. So the graph of $x\,2^{-x}$ starts out at the x-axis, climbs into the first quadrant, reaches a peak somewhere, and then falls back towards the x-axis (see Figure 11.2).

The same thing happens no matter what value a has, so long as it exceeds 1. That is, the graph of $f(x) = x\,a^{-x}$ starts out at the x-axis, climbs into the first quadrant, reaches a peak somewhere, and thereafter falls back towards the x-axis. The exact point at which the curve reaches its peak naturally depends upon the value of a.

(Note that if $a = 1$ or $0 < a < 1$ this does not happen. The curve steadily rises with increasing x, and there is no peak value reached. The curve heads determinedly towards infinity without a pause, for now the value of $x\,a^{-x}$ increases without limit as x increases without limit.)

The display above shows the graphs corresponding to $a = 2, 3$ and 4; the maxima are assumed (approximately) when $x = 1.4$, $x = 0.9$ and $x = 0.7$. Analysis shows that the exact maximizing values are as follows:

Value of a	2	3	4
Maximizing value of x	1.443	0.910	0.721

We see that the maximizing value of x exceeds 1 for $a = 2$ and slowly decreases with increasing values of a; it is less than 1 for $a = 3$ and $a = 4$.

For what value of a does it happen that the maximum value is achieved at $x = 1$? Yes, you guessed it!—*it happens precisely when $a = e$.*

11.3 Various series for e

We list here, without proof, three series associated with e. (The proofs are given in Appendix B.)

- Let n be a large positive integer; then the quantity

$$1 + \frac{1}{1!} + \frac{1}{2!} + \frac{1}{3!} + \frac{1}{4!} + \cdots + \frac{1}{n!}$$

is approximately equal to e; and the approximation improves as n increases. This result may be written alternatively as

$$e = 1 + \frac{1}{1!} + \frac{1}{2!} + \frac{1}{3!} + \frac{1}{4!} + \cdots \quad \text{(to infinity)}.$$

Here e has been written as the sum of an infinite series.

- Here is an infinite series for $1/e$:

$$\frac{1}{e} = \frac{1}{2!} - \frac{1}{3!} + \frac{1}{4!} - \frac{1}{5!} + \frac{1}{6!} - \cdots \quad \text{(to infinity)}.$$

- More generally, we have the following: for all real numbers x,

$$e^x = 1 + x + \frac{x^2}{2!} + \frac{x^3}{3!} + \cdots + \frac{x^n}{n!} + \cdots \quad \text{(to infinity)}.$$

11.4 Continued fractions for e

Here we express e as an "infinite continued fraction"—a fraction that 'goes down' indefinitely deeply into the denominator.

We first define the term "continued fraction" (CF for short). Let $a, b, c, d, \ldots$ be integers, typically positive. Then by $[a; b, c, d, \ldots]$ we shall mean the number

$$a + \cfrac{1}{b + \cfrac{1}{c + \cfrac{1}{d + \cdots}}}.$$

This is referred to as a *simple* continued fraction (SCF), as the numerators in the fraction are all 1s. If the numerators are not 1s, as in the above fraction, then the CF is known as a *general continued fraction* (GCF). For convenience, we also use the expression

$$a + \cfrac{1}{b+} \ \cfrac{1}{c+} \ \cfrac{1}{d + \cdots}$$

to denote the SCF $[a; b, c, d, \ldots]$.

It turns out that e has the following expression as an infinite CF (note that this is not a SCF):

$$e = 2 + \cfrac{3}{3 + \cfrac{4}{4 + \cfrac{5}{5 + \cdots}}},$$

Note the elegant pattern! This CF may also be written as

$$\frac{1}{e - 2} = 1 + \cfrac{1/2}{1 + \cfrac{1/3}{1 + \cfrac{1/4}{1 + \cfrac{1/5}{1 + \cdots}}}}.$$

Two other very pretty infinite continued fractions involving e are the following (the first is a SCF):

$$\frac{e^2 - 1}{e^2 + 1} = \cfrac{1}{1 + \cfrac{1}{3 + \cfrac{1}{5 + \cfrac{1}{7 + \cdots}}}},$$

and

$$\frac{1}{\sqrt{e} - 1} = 1 + \cfrac{2}{3 + \cfrac{4}{5 + \cfrac{6}{7 + \cdots}}}.$$

The following SCF for e was found by Euler:

$$e = 2 + \cfrac{1}{1+} \ \cfrac{1}{2+} \ \cfrac{1}{1+} \ \cfrac{1}{1+} \ \cfrac{1}{4+} \ \cfrac{1}{1 + \cdots} \ .$$

Here the denominators occur in an elegant repetitive pattern: 1, 2, 1, 1, 4, 1, 1, 6, 1, 1, 8, 1, ... Thus, Euler's result is

$$e = [2; 1, 2, 1, 1, 4, 1, 1, 6, 1, \ldots, \overline{1, 2k, 1}, \ldots].$$

These CFs may be used to yield the following rational approximations to e:

$$\frac{19}{7}, \quad \frac{193}{71}, \quad \frac{2721}{1001}, \quad \frac{49171}{18089}, \quad \frac{1084483}{398959}, \quad \ldots$$

Comparing these numbers with e, we find that the errors are roughly as follows:

$$4 \times 10^{-3}, \quad -3 \times 10^{-5}, \quad 1 \times 10^{-7}, \quad -3 \times 10^{-10}, \quad 5 \times 10^{-13}, \quad \ldots$$

The errors diminish at an impressive rate.

11.5 The factorial numbers

In this section we take another look at the factorial numbers which we met had above, while studying the derangement numbers (Section 11.1). Recall that if n is a positive integer, then $n!$ is the product of $1 \times 2 \times 3 \times \cdots \times (n-1) \times n$. It is easy to anticipate that the factorial numbers grow very rapidly, and computations show that this is so; thus, for example, $10! = 3628800$, $20! = 2432902008176640000$, $50!$ is a 65-digit number, and $100!$ is a 158-digit number. Whew!

This being so, let us look for ways to cut the factorial numbers down to size; bring them "down to Earth", so to speak. A simple way suggests itself—simply take the nth roots; that is, consider the numbers $a_1, a_2, a_3, \ldots$ where a_n is given by $a_n = (n!)^{1/n}$. We do just this, below. The following table gives a few values of the a-sequence.

n	20	40	60	80	100
a_n	8.30	15.77	23.19	30.60	37.99

Here are a few more such values:

$$a_{500} \approx 185.43, \quad a_{1000} \approx 369.49.$$

The graphical display in Figure 11.3 gives us a visual feel of how the a-numbers grow.

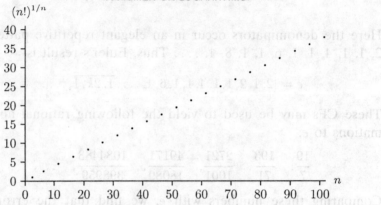

Figure 11.3. *Graph of* $(n!)^{1/n}$

We see that even after taking the nth roots, the resulting numbers grow at a steady rate—but not too rapidly; indeed, they seem to be growing in an approximately linear manner. That is, the points seem to lie roughly in a straight line. What is the slope of this line, i.e., at what rate does it climb? To check this, we compute the quantities b_n given by $b_n = a_n/n$. We get the following figures:

n	20	40	60	80	100
b_n	0.415	0.394	0.387	0.382	0.380

How curious—the numbers seems to be levelling out (i.e., they seem to approach some constant)! Further computations confirm this suspicion; we find that

$$b_{500} \approx 0.371, \quad b_{1000} \approx 0.369.$$

Now this is *very* curious—we seem to be meeting our old friend e once again, for $1/e$ is approximately 0.368. It begins to appear, then, that $b_n \approx 1/e$ when n is large; the same number once again! It had appeared before in the analysis of the derangement numbers, and now here it is again. So here is our finding: *When n is large, we have $b_n \approx 1/e$.* Recalling the definition of b_n we write this as follows:

$$\text{When } n \text{ is large, } \quad (n!)^{1/n} \approx \frac{n}{e}.$$

We may be tempted at this stage to take the nth powers on both sides and write: *When n is large, $n! \approx (n/e)^n$.* But now we find something surprising—the quantities $n!$ and $(n/e)^n$ are less close than we might expect. For instance we have, with $n = 50$ and $n = 100$,

$$50! \approx 3.041 \times 10^{64}, \qquad (50/e)^{50} \approx 1.714 \times 10^{63},$$
$$100! \approx 9.333 \times 10^{157}, \quad (100/e)^{100} \approx 3.7201 \times 10^{156}.$$

For both values of n the two computed quantities have similar magnitudes but, on the other hand, in each case, there *is* a significant difference between them.

On reflection, though, we should have anticipated this; for the relation $(n!)^{1/n} \approx n/e$ is only approximately true, and when n^{th} powers are taken, the errors are bound to accumulate in a significant manner. So there is bound to be a significant difference between $n!$ and $(n/e)^n$.

The investigation may seem to have got blocked now, but we have still more surprises in store for us. Let us look more closely at the quantity $n! \div (n/e)^n$ for large n. Write

$$c_n = \frac{n!}{(n/e)^n}.$$

The following table gives a few representative values of c_n:

n	20	40	60	80	100
c_n	11.26	15.89	19.44	22.44	25.09

Observe that c_n grows steadily as n increases, but not too rapidly. A deeper analysis, requiring the use of more advanced material (specifically, the calculus), shows that c_n is approximately proportional to $\sqrt{n}$; in fact, $c_n \approx \sqrt{2\pi n}$ for large n. Here π is the "other" mysterious number of mathematics, the other number which enjoys centre-stage status in mathematics—a number more mysterious even than e, and known since Greek times as the ratio of the circumference to the diameter of a circle, $\pi \approx 3.14159$. Good gracious! The two numbers e and π are now joining hands!

For comparison, we have, with $n = 50$, 100 and 500:

$$c_{50} \approx 17.7541, \qquad \sqrt{100\,\pi} \approx 17.7245,$$
$$c_{100} \approx 25.0872, \qquad \sqrt{200\,\pi} \approx 25.0663,$$
$$c_{500} \approx 56.0593, \qquad \sqrt{1000\,\pi} \approx 56.0499.$$

The closeness of c_n and $\sqrt{2\pi n}$ should be quite apparent now.

Tying all the threads together, we have our final glorious result:

$$\text{When } n \text{ is large,} \quad n! \approx \sqrt{2\pi n}\left(\frac{n}{e}\right)^n.$$

This result is known to mathematicians as *Stirling's formula*.

11.6 Prime numbers

We now come to the one of the most astonishing phenomena in all of mathematics—the connection between prime numbers and natural logarithms.

A positive integer n is said to be prime if it exceeds 1 and its only divisors are 1 and n. Here is the sequence of primes:

$$2, \; 3, \; 5, \; 7, \; 11, \; 13, \; 17, \; 19, \; 23, \; 29, \; 31, \; \ldots.$$

It may be shown that every positive integer is *uniquely expressible* as a product of primes. That is, we may write every positive integer n as a product of finitely many primes,

$$n = p^a \times q^b \times r^c \times \cdots,$$

where $p, q, r, \ldots$ are prime numbers and $a, b, c, \ldots$ are positive integers, and there is precisely one way of doing this. This statement is known as the *Fundamental Theorem of Arithmetic*.[1]

EXAMPLE. For $n = 100$ and $n = 221$ we have: $100 = 2^2 \times 5^2$, $221 = 13 \times 17$.

So the primes are the 'multiplicative building blocks' of the world of integers. Immediately a question arises: *Are there finitely many or infinitely many primes?* That is, does the sequence of primes come to an end? Long ago, the Greeks showed by means of a simple and elegant idea that there must be infinitely many primes.

The prime number sequence may look quite innocuous, but a lot of mystery is packed into it! There are numerous questions about this sequence which are unanswered even as of today, despite furious research on them by the world's top mathematicians. An example of such a question is: *Are there infinitely many instances of pairs of primes that differ by 2?* Some examples of such pairs are:

---

[1]Note that 1 is not considered as prime. Number theorists call it, instead, a "unit".

$(3, 5)$, $(5, 7)$, $(11, 13)$, $(17, 19)$ and $(29, 31)$. This is the notorious "twin primes" problem, and it is considered to be very difficult.

The concerted efforts to find patterns in the prime number sequence have, on the whole, met with failure. It has repeatedly happened that patterns are discovered and then found to peter out; that is, they do not persist. Efforts to find formulas that generate the primes have been given up; so also for formulas that generate only primes. Here is a quote from the great mathematician Leonhard Euler on this subject:

> Mathematicians have tried in vain to this day to discover some order in the sequence of prime numbers, and we have reason to believe that it is a mystery into which the human mind will never penetrate. To convince ourselves, we only have to glance at tables of primes, ... and we should perceive at once that there reigns neither order nor rule.

Instead, mathematicians now ask for *approximate formulas*, and one such formula will be the focus of discussion in this section.

Let p_n denote the nth prime: $p_1 = 2, p_2 = 3, p_3 = 5$, $p_{100} = 541$, ..., and let $P(x)$ denote, for positive numbers x, the number of primes less than or equal to x:

$$P(2) = 1, \ P(3) = 2, \ P(10) = 4, \ P(100) = 25, \ ...,$$
$$P(1000) = 168, \ P(10000) = 1229, \ ....$$

Curiously, the function P is easier to manage than the sequence of primes itself.

It is not too hard to notice that, on the whole, the primes thin out as we move forward into the sequence. Thus, we find that there are 168 primes below 1000; among the 1000 numbers from 10,000 to 11,000 there are 106 primes; among the 1000 numbers from 100,000 to 101,000 there are 81 primes; among the 1000 numbers from 1,000,000 to 1,001,000 there are 75 primes; among the 1000 numbers from 10,000,000 to 10,001,000 there are 61 primes; and so on. The thinning out is quite apparent in these figures.

In the eighteenth and nineteenth centuries, mathematicians found formulas that described the function P in an *approximate* manner; amazingly, they involve the logarithmic function! Gauss found that for large values of x we have

$$P(x) \approx \frac{x}{\ln x},$$

and Legendre found, by playing around with the figures, that the following formula gave a better fit:

$$P(x) \approx \frac{x}{\ln x - 1.08}.$$

The following table shows the closeness of this approximation. (For convenience, we have used the symbol $L(x)$ for Legendre's function; i.e., $L(x) = x/(\ln x - 1.08)$. The 4th column gives the absolute error in the approximation, i.e., $L(x) - P(x)$, and the last column gives the percentage error.)

x	$P(x)$	$L(x)$	error	%age error
10^2	25	28	3	12.00%
10^3	168	172	4	2.38%
10^4	1229	1230	1	0.08%
10^5	9592	9585	-7	-0.07%
10^6	78498	78521	23	0.03%
10^7	664579	664978	399	0.06%

For $x = 10^8$ we get: $P(x) = 5761455$ and $L(x) = 5766786$. The absolute error is 5331 and the percentage error is about 0.0009%. The closeness is astounding! The great mathematician Abel (from Norway) once described the closeness as "the most remarkable in all mathematics".

Gauss did not stop with his discovery that $P(x) \approx x/\ln x$, and went on to find better formulas. (The approximation $P(x) \approx x/\ln x$ is actually not too impressive for small values of x; Legendre's formula gives much better results.) The most remarkable of these is the following.

We consider the graph of $\ln x$. This graph has the appearance shown in Figure 11.4.

Let $\mathrm{Li}(t)$ denote the area of the shaded region shown in the graph – i.e., the area of the region enclosed by the curve – the x-axis, and the ordinates $x = 2$ and $x = t$. Gauss found that

$$P(x) \approx \mathrm{Li}(x).$$

For comparison with Legendre's approximation, we provide a table similar to the earlier one.

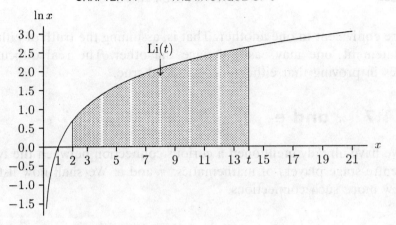

Figure 11.4. *Graph of* ln *x*

x	$P(x)$	Li(x)	error	%age error
10^2	25	30	5	20.00%
10^3	168	178	10	5.95%
10^4	1229	1246	17	1.38%
10^5	9592	9630	38	0.40%
10^6	78498	78628	130	0.17%
10^7	664579	664918	339	0.05%

For $x = 10^8$ we get: $P(x) = 5761455$ and Li$(x) = 5762209$; the absolute error is 754 and the percentage error is 0.0001%. The results certainly take one's breath away! Indeed, if anything they get more impressive for larger values of x. Thus, for $x = 10^{10}$ we get: $P(x) = 455052511$ and Li$(x) = 455055615$; the absolute error is 3104 and the percentage error is about 0.00007%!

Observe that all the study done so far is *empirical*—we have worked only with "experimental data"; no "theory" has yet entered the scene. It turns out that the underlying theory is extremely difficult, and proving that $P(x) \approx$ Li(x) is a most formidable task.

The statement that $P(x) \approx x/\ln x$ for large values of x (more precisely, that the ratio $P(x) \div x/\ln x$ is close to 1 for large values of x and *tends* to 1 as x increases without limit) is known as the *Prime Number Theorem*.

COMMENT. It is not too hard to show that the following two statements:
- $P(x)/L(x)$ tends to 1 as x increases without limit;
- $P(x)/$Li(x) tends to 1 as x increases without limit;

are equivalent to one another. That is, assuming the truth of either statement, one may easily deduce the other. The real difficulty lies in proving that either one is indeed true.

11.7 π and e

We have already mentioned a curious connection between the two centre-stage players of mathematics: π and e. We shall now list a few more such connections.

11.7.1 Euler's formula

This involves the 'imaginary' number i, which represents the square root of -1. This number is not part of the set of real numbers, so it is in some senses a piece of fiction. (Purists however may say that $\sqrt{2}$ is just as much a piece of fiction, as it does not belong the set of rational numbers.) Curiously, the introduction of this piece of 'fiction' enriches mathematics very greatly; some of the most beautiful theorems of mathematics involve i, and (most curious of all) physics makes a lot of use of i, as does electrical engineering.

By "mixing" i with the real numbers and making a sort of soup of the two, we get the numbers known as *complex numbers*; e.g., numbers such as $1 + i$, $2 - 3i$, $0.2 - 0.9i$, and so on. Note that the real numbers form a subset of the set of complex numbers.

Here is Euler's formula: for all complex numbers z,

$$e^{iz} = \cos z + i \sin z.$$

An extraordinary relation! By putting $z = \pi$ we get, since $\cos \pi = -1$ and $\sin \pi = 0$,

$$e^{\pi i} + 1 = 0.$$

There is something almost mystical about this equation, which unites five of the fundamental numbers of mathematics (1, 0, e, π and i).

11.7.2 The "bell curve"

The bell curve is well-known in statistics; many frequency distributions in nature have this appearance. The equation of the bell curve is $y = e^{-x^2}$, and the curve appears as shown in Figure 11.5.

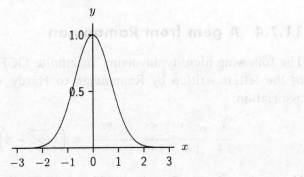

Figure 11.5. *The "bell curve"*

It is of interest to compute the area of the entire region enclosed between the curve and the x-axis. (This corresponds, in statistics, to the total frequency.) Though the region extends infinitely far along the x-axis (for e^{-x^2} never vanishes, no matter how large x is), it thins out at great speed, and the area is a finite number. Surprise! The area turns out to be equal to $\sqrt{\pi}$.

11.7.3   An infinite series plus an infinite GCF

Another curious relation is the sum of the infinite series

$$1 + \frac{1}{1 \times 3} + \frac{1}{1 \times 3 \times 5} + \frac{1}{1 \times 3 \times 5 \times 7} + \cdots,$$

and the infinite GCF (general continued fraction)

$$\frac{1}{1+} \; \frac{1}{1+} \; \frac{2}{1+} \; \frac{3}{1+} \; \frac{4}{1+ \cdots},$$

which may be written in expanded form as

$$\cfrac{1}{1 + \cfrac{1}{1 + \cfrac{2}{1 + \cfrac{3}{1 + \cfrac{4}{1 + \cdots}}}}}.$$

It turns out that the sum is the number $\sqrt{\pi e/2}$. (Whew!)

11.7.4 A gem from Ramanujan

The following identity, involving an infinite GCF, appeared in one of the letters written by Ramanujan to Hardy, during their early association:

$$\frac{1}{1+}\ \frac{e^{-2\pi}}{1+}\ \frac{e^{-4\pi}}{1+}\ \frac{e^{-6\pi}}{1+\cdots} = \left(\sqrt{5\phi} - \phi\right) e^{2\pi/5}.$$

Here $\phi = (\sqrt{5}+1)/2$ is the "golden ratio"—another centre-stage number of mathematics!

11.8 A final comment

The reader may have observed that very little has actually been *proved* in this chapter. Rather, the occurrences of e within different areas of mathematics and in different problems have been *empirically* found. This means that much of the content of this chapter may be classified as "experimental mathematics", making it akin to experimental physics. However, no mathematician is willing to stop with empirical discoveries; he/she demands proof, and will accept nothing less—certainly nothing that pretends to be a proof but actually isn't.[2] The reader may wonder about the proofs of the various statements made in this chapter. The reasons for their non-inclusion are, however, not all the same.

In the problem about derangements, the elementary principles of counting ("permutations and combinations"), in particular the result known as the "principle of inclusion-exclusion", may be used to show that D_n is the integer closest to $n!/e$. In the case of Stirling's formula,

$$n! \approx \sqrt{2\pi n}\,(n/e)^n,$$

the proof comes from estimating the area under the logarithmic curve ($y = \ln x$). Though the argument needs delicate handling (the error term in particular), it is now part of any undergraduate level course in calculus. Euler's relation ($e^{iz} = \cos z + i \sin z$) is now standard fare in any college-level course, as is the proof that the area under the bell curve is $\sqrt{\pi}$.

---

[2] *An amusing interlude.* For a physicist, all odd numbers greater than 1 are prime; for 3, 5 and 7 are prime, as are 11 and 13; in the case of 9 there is clearly some experimental error!

In the case of the prime number theorem, however, the level of difficulty goes up by several levels. The proof is very difficult and demands deep application of the theory of complex variables. Some idea of the difficulty may be gauged from the following: after Gauss found empirically that $P(x) \approx \mathrm{Li}(x)$ for large values of x, the theoretical justification of this took about three-quarters of a century (after furious assaults by the world's top mathematicians). Gauss himself did not succeed in proving the result, and it was only after Riemann, another great mathematician, had opened up completely new pathways into the problem did mathematicians find success (it came in the last decade of the nineteenth century).

In the middle of the twentieth century, two mathematicians, Paul Erdös and Atle Selberg, independently found another proof of the prime number theorem. Their proof is termed as 'elementary'. The adjective 'elementary' has a purely technical significance here, and its only connotation is that the proof uses nothing from the theory of complex variables. This proof came as a very great surprise, as it was by then widely accepted that one could not do without complex variables. Note that there is nothing *easy* about the proof; if anything, the non-elementary proof – difficult though it is – is easier to grasp than the elementary proof!

Ramanujan's identity belongs to a very advanced branch of mathematics and has, to the best of the author's knowledge, no elementary proof.

EXERCISES

11.8.1 We found earlier that the sequences $\{a_n\}$ and $\{b_n\}$ defined by

$$a_n = \left(1 + \frac{1}{n}\right)^n, \quad b_n = \left(1 + \frac{1}{n}\right)^{n+1},$$

both level off at e, but from the opposite direction; $\{a_n\}$ is monotonic increasing, whereas $\{b_n\}$ is monotonic decreasing.

Investigate numerically the behaviour of the sequences $\{u_n\}$ and $\{v_n\}$ defined as follows:

$$u_n = \left(1 + \frac{1}{n}\right)^{n+1/4}, \quad v_n = \left(1 + \frac{1}{n}\right)^{n+1/2}.$$

11.8.2 Use the fact that

$$e = 1 + \frac{1}{1!} + \frac{1}{2!} + \frac{1}{3!} + \frac{1}{4!} + \cdots \quad \text{(to infinity)}$$

to show that e is an irrational number; that is, e cannot be written as the ratio of two non-zero numbers.

Chapter 12

Miscellaneous Applications

In this chapter, we shall consider various areas of application of the logarithmic and exponential functions—areas where the two functions make what may seem at first sight to be a sudden and very unexpected appearance.

Actually, there is nothing magical or mysterious about the entry of these functions. In most instances, the reason for their appearance is simple and understandable and has to do with their special nature and properties. These properties often match exactly with certain properties of natural phenomena; hence the 'marriage' of the two.[1]

We can make this more explicit by the following description. Let $y = f(x)$ be a function of x with the property that the rate of growth of y with respect to x at any particular point is *proportional to the value of y at that point*. That is, the slope of the curve $y = f(x)$ at any point on the curve is proportional to the height (the y-value) at that point. (This happens, say, when y refers to population and x to time; for the rate of growth of population is clearly proportional to the population itself.) In this case, we can say for certain that y is an exponential function; that is, $y = a e^{bx}$ for some constants a and b. Similarly, *if the slope is inversely proportional to the x-value*, then we can say for certain that y is a logarithmic function; that is, $y = a \ln x$ for some constant a.

In practice, these conditions may hold approximately rather than exactly, and in such cases the exp and log functions may

---

[1] *Comment.* Perhaps what is really mysterious is the "why" of the phenomenon itself. Why should so many natural phenomena have just the right properties so that they can be modeled by exponential and logarithmic functions?

not be the most appropriate ones to use; alternatives may be available. Below, some mention is made of these issues. All told, however, it is remarkable how widely these two functions are used in the sciences.

12.1   Radioactivity

Radioactivity (definition) The spontaneous emission of energy from an atomic nucleus in the form of particles or radiation or both.

EXAMPLE. (a) The emission of α rays (helium nuclei) from Uranium-238, changing it into Uranium-234; (b) the emission of α rays from Radium-226, changing it into Radon-222; and so on.

Radioactivity offers a nice example of a natural phenomenon, where exponential and logarithmic functions enter the scene in a most natural fashion.

Radioactivity was discovered in the closing years of the nineteenth century by Antoine Bequerel, just as some physicists were rashly concluding that everything of significance in physics had already been discovered, with only footnotes and marginal comments left to be added, and that new talents would be well-advised to seek their fortunes elsewhere. It is hard to come across a better example of a rash prediction! With the dawn of the new century, a brand new chapter was opened in physics, and it brought to the center-stage researchers such as the Curies, Einstein, Bohr, J J Thomson, Rutherford, Schrödinger, Heisenberg, Born, Dirac, and a great many others—some of the greatest figures known to the world of physics. The period 1900–1930 must surely rank as one of the greatest and most exciting eras known in the history of science.

We shall not dwell on the historical development of the subject, nor on its many fascinating details (α rays, β rays, γ rays, ...); many excellent books are available for this purpose, and the student has, in all likelihood, studied the material as part of the school physics curriculum. We shall limit ourselves to two observations:

- Some elements are more radioactive than others; e.g., polonium and radium are more radioactive than uranium, and correspondingly, more dangerous.

- The differences in rates of decay can be exploited to yield highly reliable methods for dating fossils and ancient rock layers.

It is found that each individual radioactive element has its own characteristic rate of decay, which remains the same under all conditions and at all locations; it is intrinsic to the element itself (indeed, to the isotope: uranium-235 and uranium-238 have different decay rates). Moreover, the decay process is such that the number of atoms that decay in a given unit of time always bears the same ratio to the number of undecayed atoms. The ratio is the *decay constant* of the element and it is usually designated by the Greek letter λ (pronounced 'lambda'). Thus, each element has its individual λ. Another number which carries the same information as the decay rate is the *half-life*, t_H; this is the time taken for the number of undecayed atoms to diminish by one-half. Clearly, the larger the decay constant, the smaller the half-life. (We shall shortly find a relation between λ and t_H.) Half-lives vary greatly: from less than a second to more than a billion years. Table 12.1 gives some representative data (Z refers to the *atomic number* of the element—i.e., the number of protons in the nucleus). The enormous range in half-life is evident from the table.

Earlier, we had defined the radioactive decay constant as the ratio of the number of atoms that decay in a given unit of time to the number of undecayed atoms. Each radioactive substance has its own characteristic decay constant, which is a measure of its level of radioactivity; the higher the λ, the more radioactive the substance. The level may also be measured in *Curies*, one Curie being the number of disintegrations occurring in 1 gram of radium per second (1 Curie is equivalent to 3.7×10^{10} dps (disintegrations per second)), in *millicuries* (1/1000 of a curie or 3.7×10^7 dps), or in *microcuries* (1/1000000 of a curie or 3.7×10^4 dps).

Relation between decay constant and half-life We now derive a simple relationship between the decay constant λ and the half-life t_H. Let N_0 be the number of (undecayed) radioactive atoms present at the start ($t = 0$), and let N_t refer to the number of undecayed atoms at time t. The relationship between N_t and t is an exponential one, namely

$$N_t = N_0\, e^{-\lambda t}.$$

Table 12.1 *Half-lives of some radioactive substances*

Z	Symbol	Name	Decay	Half-life
1	n	Neutron	β	10.8 minutes
1	H	Tritium-3	β	12.3 years
2	He	Helium-5	β	6×10^{-20} seconds
6	C	Carbon-14	β	5570 years
15	K	Phosphorum-32	β	14.5 days
19	K	Potassium-42	β	12.5 hours
27	Co	Cobalt-60	γ	5.29 years
38	Sr	Strontium-88	β	51 days
38	Sr	Strontium-90	β	28 years
47	Ag	Silver-107	β	7.5 days
49	In	Indium-115	β	6×10^{14} years
53	I	Iodine-128	β	25 minutes
53	I	Iodine-131	β	8.1 days
55	Cs	Caesium-137	β	28 years
56	Ba	Barium-139	β	85 minutes
56	Ba	Barium-140	β	12.8 days
61	Pm	Promethium-147	β	2.6 years
82	Pb	Lead-209	β	3.3 hours
84	Po	Polonium-209	α	200 years
85	At	Astatine-210	α	8.3 hours
86	Rn	Radon-220	α	52 seconds
86	Rn	Radon-222	α	3.83 days
87	Fr	Francium-223	β	22 minutes
88	Ra	Radium-226	α	1600 years
89	Ac	Actinium-227	β	22 years
90	Th	Thorium-232	α	1.41×10^{10} years
91	Pa	Protoactinum-231	α	3.4×10^4 years
92	U	Uranium-235	α	7.1×10^8 years
92	U	Uranium-238	α	4.5×10^9 years
94	Pu	Plutonium-239	α	2.4×10^4 years

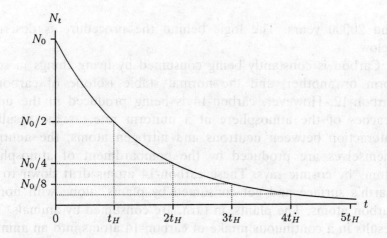

Figure 12.1. *Graph of N_t*

Table 12.2 *Typical values for $1/\lambda$*

Isotope	Half-life (t_H)	$1/\lambda$
Uranium-238	4.5×10^9 years	5.0×10^{-18}
Plutonium-239	2.4×10^4 years	9.2×10^{-13}
Carbon-14	5570 years	3.9×10^{-12}
Radium-226	1622 years	1.35×10^{-11}
Radon-222	31.1 minutes	3.7×10^{-4}
Bismuth-214	1.6×10^{-4} seconds	4.33×10^3
Helium-5	6×10^{-20} seconds	1.2×10^{19}

The graph of N_t against t is shown in Figure 12.1.

By definition, the number of undecayed atoms at time $t = t_H$ is half that at time 0; so $N_t = N_0/2$ when $t = t_H$, $\therefore e^{-\lambda t_H} = 1/2$, which means that $e^{\lambda t_H} = 2$. It follows that $\lambda t_H = \ln 2$; i.e.,

$$t_H = \frac{\ln 2}{\lambda} \approx \frac{0.693}{\lambda}.$$

So t_H is inversely proportional to λ.

Table 12.2 gives some typical values of $1/\lambda$.

Carbon dating An extremely important use to which such findings are put is in the dating of objects—fossils, rocks, the Shroud of Turin, and so on. Of particular significance is the usage of carbon-14 in dating. Since the half-life of carbon-14 is 5570 years, this turns out to be particularly useful for dating archaeological specimens, as the "typical" age of such artefacts lies between 1000 and 20000 years. The logic behind the procedure is described below.

Carbon is constantly being consumed by living things in some form or another, and the normal stable isotope of carbon is carbon-12. However, carbon-14 is being produced in the upper reaches of the atmosphere at a uniform rate, as the result of interaction between neutrons and nitrogen atoms; the neutrons themselves are produced by the bombardment of atmospheric atoms by cosmic rays. These carbon-14 atoms drift down to the Earth's surface and get consumed by plants, along with normal carbon atoms. The plants, in turn, are consumed by animals. This results in a continuous intake of carbon-14 atoms into an animal's body as long as it is alive; so the proportion of carbon-14 atoms in the body remains at a fairly constant level. However, once the animal dies its intake of carbon ceases; the decay of carbon-14 now causes the proportion of carbon-12 to carbon-14 to increase at a steady rate. So, by measuring the proportion of carbon-12 to carbon-14, we are able to estimate the number of years that have elapsed since the animal has died.

It should be clear that this method may be used not only to date animal remains but also human artefacts made of material such as wood, cloth or paper. (This explains the reference made above to the Shroud of Turin. However, there continues to be considerable controversy surrounding this matter, and we shall not say any more.)

Underlying assumption of carbon dating technique An obvious weakness of the carbon-dating technique is that it assumes implicitly that the rate of production of carbon-14 has remained constant since distant times. This, in turn, requires that the intensity of interstellar cosmic rays has remained constant, and this is quite possibly not true.

12.2 Population growth

Every society has associated with it a *birth rate*, which remains fairly constant with time, changing only when there are radical changes in the societal structure. This implies that the increase in population that happens in a fixed interval of time, say a year, *increases* with time. The data in Table 12.3 bear this out (population figures are given in units of one billion; so the world population in the year 1700 was about 0.63 billion or 630 million; and so on). A graphical display is shown too (Figure 12.2).

Table 12.3 *Data on population*

t	P_t	t	P_t	t	P_t
1650	0.51	1850	1.13	1970	3.72
1700	0.63	1900	1.60	1980	4.48
1750	0.71	1950	2.57	1990	5.32
1800	0.91	1960	3.05	2000	6.05

The following equation, connecting the population P_t in year t with time t, provides a reasonably good model of population growth:

$$P_t = P_0 e^{bt}.$$

Here P_0 is the population in year 0 (any convenient year may be designated as year 0), t is the number of years that have elapsed since year 0, and b is the birth rate. This equation is the logical and inevitable result of assuming that the rate of growth in year t is proportional to the population in that year; it predicts that population increases at an exponential rate. A "population explosion" is clearly going to arise within this model.

A more realistic model for the growth of population Does the model described above really fit the facts? There are difficulties; as long as food is available in plenty, population is bound to grow exponentially, but once food limitations enter the picture the

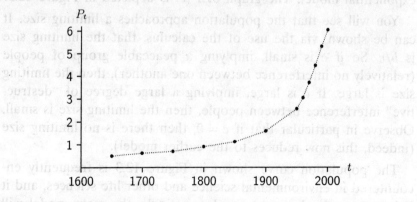

Figure 12.2. *Population–time graph*

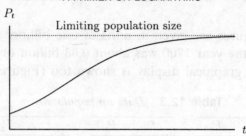

Figure 12.3. *Limits to population growth*

profile changes in a significant manner. (This may be observed in a laboratory setting, working with bacteria growing in a vessel where food is supplied at a constant rate. In the beginning, the rate of growth is as predicted by the exponential model, but as the population grows the rate of growth is seen to wear off.)

It has been proposed that a better model for population growth is obtained by taking the rate of growth to be of the form

$$bP - cP^2 \qquad \text{(rather than just } bP\text{).}$$

Here c is a positive constant and the term cP^2 represents the interaction between people. We may take it to represent the effects due to competition—as P rises, the quantity cP^2 increases, thus reducing the rate of growth of P; it 'pulls' down the population growth. If this model is carefully analyzed, its prediction for population growth turns out to be very different from that given by the exponential model. The graph of $P(t)$ is depicted in Figure 12.3.

You will see that the population approaches a limiting size. It can be shown, via the use of the calculus, that the limiting size is b/c. So if c is small, implying a peaceable group of people (relatively no interference between one another), then the limiting size is large. If c is large, implying a large degree of "destructive" interference between people, then the limiting size is small. Observe in particular that if $c = 0$, then there is no limiting size (indeed, this now reduces to the earlier model).

The population curve shown in Figure 12.3 is frequently encountered in environmental science and other life sciences, and it is known as the *logistic curve*. Interestingly, the same model will serve to describe the growth of a rumour!

12.3 Cooling down

As is commonly observed, a hot cup of coffee placed on a table in a room gradually cools down, and the rate at which it cools down is not a constant—the hotter the coffee, the faster it cools down. We observe likewise that a cold cup of coffee warms up, and the colder it is, the faster it gains in temperature. In short, the tendency for a hot or cold body to assume the ambient temperature (= the temperature of the surroundings) depends upon the difference in temperature—the larger the difference, the faster the rate. Isaac Newton proposed the law that now bears his name: *Newton's law of cooling*. This may be stated as follows:

The rate of loss of heat of a hot body when placed in a steady stream of air is proportional to the difference between its temperature and that of the surroundings.

Since the heat content of a body is $C \times \theta$, where C is its thermal capacity (a fixed constant for any given body) and θ is its temperature, the word 'heat' in the above law may be replaced by 'temperature' So Newton's law of cooling may also be written as

The rate of change of the temperature θ of a hot body placed in a steady stream of air is proportional to $\theta - \theta_s$, where θ_s is the ambient temperature.

Note that the constant of proportionality must be negative, because a hot body cools down, whereas a cold body warms up.

If x stands for $\theta - \theta_s$, then Newton's law implies that the rate of change of x with time is proportional to x, implying as earlier that x is an exponential function of the time t. This leads to the relation

$$\theta = \theta_s + (\theta_0 - \theta_s) e^{-kt},$$

where θ_0 is the temperature of the body at time 0, and k is a positive constant whose value depends upon various physical characteristics of the body: thermal capacity, colour, shape, surface area, and so on. (As is well known, darker colours and larger surface area lead to faster heat loss.)

The variation of temperature with time is best seen via a graph; see the display in Figure 12.4.

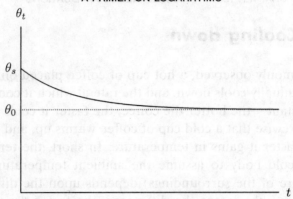

Figure 12.4. *Cooling down (forced convection)*

How well does Newton's law fit the facts? Newton's law has been experimentally found to fit the facts well when there is a steady stream of air over the body (conditions of "forced convection"), particularly when the speed of the air is in excess of 4 m/s. However, when the air is still (conditions of "natural convection") the following law is experimentally found to be the more appropriate one:

> *The rate of heat loss of a body under conditions of natural convection is proportional to the 5/4 power of the difference between the temperature of the body and that of its surroundings.*

So if x represents the difference between the temperature of the body and the ambient temperature, then the rate of change of x with respect to time is proportional to $x^{5/4}$. When this law is analyzed mathematically, it is found that the exponential relation vanishes! Instead, the relation between θ and t is the following:

$$\theta = \theta_s + \frac{1}{(b+kt)^4},$$

where k has the same significance as earlier, and b depends on the difference between the initial temperature of the body and the ambient temperature (in fact, $b = (\theta_0 - \theta_s)^{-1/4}$). The variation of θ with t is shown in Figure 12.5. The similarity with the earlier situation is obvious.

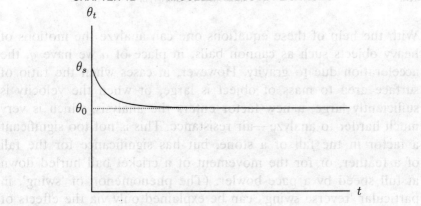

Figure 12.5. *Cooling down* (*natural convection*)

12.4 Parachuting down

Legend has it that the great Italian physicist Galileo Galilei once dropped two cannon balls of different sizes from the top of the famous Leaning Tower of Pisa (which actually does lean!), to check the ancient and unquestioned wisdom that heavy objects fall more rapidly than light objects. Of course, they fell at exactly the same rate, and so a new discovery in science was made. Whether Galileo actually did perform such an experiment is uncertain, but he certainly did make a very careful study of falling bodies, and his work marks a very important milestone in the history of science, for he established facts that went against the ancient wisdom passed down since Greek times. Most importantly, he established the modern approach to science, based on experiment and mathematical analysis. The reader may wonder at the basis of the ancient wisdom. Perhaps it lies in the common observation that stones and feathers do not in fact fall at the same rate; stones fall a great deal faster. Since feathers are much lighter than stones, can one not conclude that heavier objects fall more rapidly? It is a tribute to Galileo's genius that he did not jump to this "obvious" conclusion.

A stone falling from the top of a building falls according to the well-known equations of motion: if a body is moving in a straight line with uniform acceleration a, and in a span of time t it moves through a distance s, its initial and final velocities being u and v respectively, then

$$v = u + at, \quad s = ut + \tfrac{1}{2}at^2, \quad v^2 - u^2 = 2as.$$

With the help of these equations one can analyze the motions of heavy objects such as cannon balls; in place of a we have g, the acceleration due to gravity. However, in cases when the ratio of surface area to mass of object is large, or when the velocity is sufficiently large, a new factor enters the analysis, which is very much harder to analyze—air resistance. This is not too significant a factor in the fall of a stone, but has significance for the fall of a feather, or for the movement of a cricket ball hurled down at full speed by a pace bowler. (The phenomenon of 'swing', in particular 'reverse swing', can be explained only via the effects of air resistance.)

Notation   We shall use the following notation in our analysis: if x is a quantity that varies with time, then x' denotes the *rate* at which it varies. So if x is a displacement (i.e., distance travelled by a moving body), then x' will denote velocity; and if x is a velocity, then x' will denote acceleration. We may, if we wish, use two dashes; so if x denotes displacement, then x'' will denote acceleration.

★ ★ ★

Let a body with mass m fall from a state of rest at height h. At any intermediate point in its fall, let s, v and a represent (respectively) its displacement from the starting point, velocity and acceleration. Then,

$$v = s'; \quad a = v' = s''; \quad \text{at the start } s = 0 \text{ and } v = 0.$$

In the analysis below we make repeated use of Newton's first law of motion (*force equals mass times acceleration*, or $F = ma$) and the law of gravity (*force exerted by Earth on the object equals* mg).

If there is no air resistance, then we have, equating forces,

$$ma = mg, \quad \therefore \ s'' = g, \quad \therefore \ v' = g.$$

This kind of equation, involving rates of change, is referred to in higher mathematics as a *differential equation*. This particular differential equation is very easy to solve: we get $v = gt$ and $s = \frac{1}{2}gt^2$, and these equations are identical to those given earlier, with the initial velocity u equal to 0.

★ ★ ★

Suppose, however, that there is air resistance. Unfortunately, the problem can now change from being very simple to being very complex. The difficulty is that air resistance is an extremely complicated phenomenon, and its laws are not easy to fathom. Evidence for this statement can be gauged from the fact that aeronautical engineering is one of the most difficult branches of engineering! The point is that air resistance depends upon a very large number of factors: the speed of the object (the faster the object moves, the larger the air resistance; further, the equation connecting the two may itself be different at different speeds); its shape (it is a matter of common observation that shape influences wind resistance very greatly; minute changes in shape can result in large differences in air resistance. Hence, the need to test new wing designs, car designs, etc. in devices known as wind tunnels, to experimentally measure the air resistance, this being easier and more reliable than a theoretical calculation); texture of its surface; viscosity of the medium; and so on.

Resistance proportional to the velocity There are some cases, however, when the problem is fairly easy to solve: when the air resistance is a "nice" function of the velocity of the object (in the sense of being mathematically easy to handle); in particular for the case when the air resistance is proportional to the velocity. That is, the frictional force due to air is of the form kv for some positive constant k. (Here k must have the units of mass/time.) Since the frictional force acts against the direction of motion, we get, by Newton's first law,

$$ma = mg - kv, \quad \therefore \quad mv' = mg - kv, \quad \therefore \quad v' = g - bv,$$

where $b = k/m$ is a positive constant (its units are 1/time). This too is a differential equation (but of another kind), and techniques are available for its solution. We get, after some manipulation,

$$v = \frac{g}{b}\left(1 - e^{-bt}\right).$$

So the solution involves the exponential function. Observe that the quantity in the exponent, bt, is dimensionless (its units are 1/time × time).

The graph of v against time t is shown in Figure 12.6; we see that v behaves in much the same way as the temperature of a cold

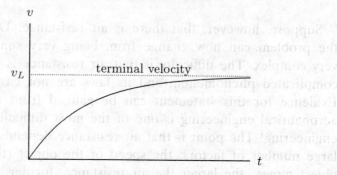

Figure 12.6. *Air resistance proportional to speed*

body warming up. As t increases without limit, the quantity e^{-bt} diminishes rapidly, and the velocity approaches a limiting value v_L, given by $v_L = g/b$. This is the *terminal velocity* of the object. Once the body attains this velocity in a state of free fall, it does not accelerate any further, because the forces of gravitation and frictional air resistance exactly balance out; and the status quo continues thereafter. This will naturally happen for any object falling under gravity in any fluid. Thus, a ball bearing falling in a cylindrical column of glycerine will attain a limiting velocity soon enough.

Any object has associated with it a terminal velocity; this is its limiting speed in a state of free fall in the atmosphere of the Earth. The terminal velocity depends upon many factors: the mass of the object, its shape, its texture (i.e., whether its surface is rough or smooth), and so on. Thus, the terminal velocity for an ant or a feather is extremely small, and it is rather small for a mouse too. There is an amusing consequence of this: a mouse that falls off a skyscraper, or from a plane, does not get hurt on landing. The reader may have observed a squirrel fall from a tree during its frolic. On landing it looks a bit dazed, but only for a few moments; in no time at all it has scrambled up back to the tree, quite unhurt! Unfortunately for us, the terminal velocity for a human being is uncomfortably high—large enough to kill us, in fact. Parachutes are designed to reduce the terminal speed to a level that we can handle (but training is needed to cope even with this vastly reduced speed). The terminal velocity of a raindrop is low, and that of a fine droplet of oil is exceedingly small; this means that the droplets take an extremely long time to settle down.

COMMENT. If an object is fired downwards, from a height, with a starting velocity that exceeds its terminal velocity v_L, then it will in fact *decelerate*. Eventually, its velocity will reach the terminal velocity.

Resistance proportional to square of velocity   The equation $bv = g(1-e^{-bt})$ was obtained assuming that air resistance is proportional to velocity. But this may not be the case; the dependence of air resistance on velocity may be more complicated. One possibility is that that air resistance depends on the *square* of the velocity, say it is equal to cv^2 where $c > 0$ is some constant (its dimensions are mass/length). In this case, we get

$$mv' = mg - cv^2, \quad \therefore \ v' = g - \frac{c}{m}v^2 = g - dv^2, \text{ say,}$$

where $d = c/m$ is a constant (its units are 1/length). This yet again is a differential equation and it too may be solved using standard methods.

Write v_L for $\sqrt{g/d}$ and note that v_L has the units of length/time, which means that it represents a velocity. Also write α for $2bv_L$, then α is a quantity with units 1/time, therefore αt is a dimensionless quantity. Analysis now shows that

$$v = v_L \left(\frac{e^{\alpha t} - 1}{e^{\alpha t} + 1} \right).$$

From this equation, we can directly see that v_L is in fact the terminal velocity of the object. For as t increases without limit, the quantity $e^{\alpha t}$ increases without limit, and so the quantity within brackets gets closer and closer to 1 (the '1' is insignificant in comparison to $e^{\alpha t}$, and so the fraction is of the form $e^{\alpha t}/e^{\alpha t}$). Thus, the terminal velocity in this case is given by $v_L = \sqrt{g/d} = \sqrt{mg/c}$.

The variation of v with time t is shown in Figure 12.7. It appears from the graphs that it takes longer for the terminal speed to be reached in the second situation (which makes sense).

★ ★ ★

Finally, here is a tricky problem for readers to mull over. Let an object be thrown vertically upwards from the ground at a speed of v. In the absence of air resistance, the time taken to reach the maximum height would be v/g, $\therefore$ the time taken to return

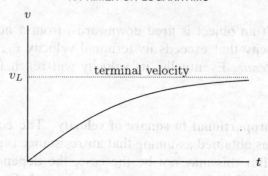

Figure 12.7. *Air resistance proportional to the square of speed*

to ground level would be $2v/g$. What does the resistance of air do? Obviously, velocity gets reduced, as does the maximum height reached. Question: *Is the total travel time less or more when there is air resistance?*

12.5 The hanging necklace

Borrow a necklace and hold it up in the air by its ends, with both ends at the same level, so that it hangs downwards. The necklace assumes the shape of a curve, symmetrical about its lowest point. The curve is called a *catenary*, from the Latin word for chain. Question: *What kind of curve is a catenary? What is its equation?* This problem was first posed by Jakob Bernoulli (1654–1705), one of the many members of the illustrious (and amazing, but also cantankerous!) Bernoulli clan. He circulated the following problem as a challenge to the leading mathematicians of the day: *[Find] the curve assumed by a loose string hung freely from two fixed points.* (It is assumed that the string has uniform density all along its length and is completely flexible.) At first sight it may seem as though the curve is a parabola, but this is not so. Determining the equation of the catenary, however, requires a good degree of familiarity with both physics and differential equations, and we shall not go too deeply into the analysis of the problem.

Bernoulli's challenge drew three different solutions—by Christian Huygens (1629–1695; physicist, astronomer and mathematician, and a contemporary of Newton's), Gottfried Leibnitz (1646–1716; mathematician, philosopher and many more things; and co-creator, along with Newton, of the calculus), and Johann Bernoulli

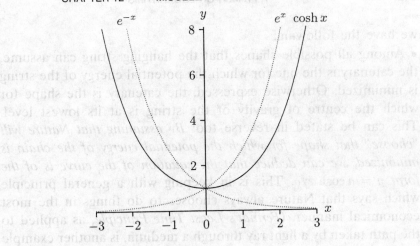

Figure 12.8. *The shape of a hanging chain*

(1667–1748), another member of the Bernoulli clan. Each one analyzed the problem in a different way, but their conclusions were of course the same. Here is what they showed.

Let the x-axis be a horizontal line in the plane of the curve, and let the y-axis be the vertical line of symmetry of the curve. Then, the equation of the catenary has the form

$$y = a\left(\frac{e^{x/a} + e^{-x/a}}{2}\right) + b,$$

where a and b are constants; a depends on the physical character-istics (e.g., density) of the chain and on how far apart its ends are held (a is large if the curve is narrow and deep, and a is small if the curve is wide and shallow); and b depends on the placement of the x-axis. By placing the axis a units below the lowest point of the curve (i.e., $y = a$ when $x = 0$), we may take b to be 0. The equation now takes the form $y = a(e^{x/a} + e^{-x/a})/2$. A sketch of this curve is shown in Figure 12.8.

The function $(e^x + e^{-x})/2$ often occurs in mathematics and has earned a name for itself: it is called the *hyperbolic cosine function*, or 'cosh' for short. So the equation of the catenary is $y = a \cosh x/a$. When $a = 1$, the catenary may be thought of as the "average" of the two curves $y = e^x$ and $y = e^{-x}$; see Figure 12.8.

★ ★ ★

Remarkably, the catenary arises in other contexts. For example, we have the following:

• Among all possible shapes that the hanging string can assume, the catenary is the one for which the potential energy of the string is minimized. Otherwise expressed, the catenary is the shape for which the centre of gravity of the string is at its lowest level. This can be stated in reverse too: *By assuming that Nature will "choose" that shape for which the potential energy of the chain is minimized, we can deduce that the equation of the curve is of the form* $y = a \cosh x/a$. This is in keeping with a general principle which says that Nature always chooses to do things in the most economical manner. (*Fermat's Least Time Principle*, as applied to the path taken by a light ray through a medium, is another example of this "economy".)

• Let P and Q be two fixed points lying above the x-axis of a 2-D coordinate system, and let C be any (variable) arc connecting P and Q, lying completely above the axis. Let C be rotated around the x-axis through an angle of 360°, and let Area(C) be the surface area of the resulting object. We wish to choose the arc so that Area(C) is minimized. Which curve accomplishes this? It is the catenary; we must choose the arc so that it forms part of a catenary.

Here is an interesting and exciting application of this. Dip two identical circular hoops of wire into a soap solution, in contact with one another along their entire circumference. Let the hoops be pulled out of the solution and drawn apart carefully. The soap solution produces a soap film connecting the two hoops, and the shape assumed by the film is a catenoid, the surface produced by rotating a portion of a catenary around the x-axis. We see here the optimizing principle at work once more: the soap film takes that shape which minimizes the total potential energy, and this shape is the catenoid.

Is there a hyperbolic sine too?   We introduced the "hyperbolic cosine" function above, and the reader may wonder if there is a "hyperbolic sine" too. There is; it is denoted by 'sinh' for short, and $\sinh x$ is defined as $(e^x - e^{-x})/2$. There are close similarities between the hyperbolic sine and cosine functions and the trigonometric sine and cosine functions, and many properties of the two hyperbolic functions are related to properties of the hyperbola; hence their curious names.[2]

---

[2]Pronunciation of these names tends to be a problem! Some people pronounce

Visual similarity between a catenary and a parabola Before concluding this section, let us explain the visual similarity between a catenary and a parabola. In the previous chapter, we saw that e^x could be written as the sum of an infinite series of powers of x, as follows:

$$e^x = 1 + x + \frac{x^2}{2!} + \frac{x^3}{3!} + \frac{x^4}{4!} + \cdots.$$

By writing $-x$ in place of x, we obtain a similar expression for e^{-x}

$$e^{-x} = 1 - x + \frac{x^2}{2!} - \frac{x^3}{3!} + \frac{x^4}{4!} - \cdots.$$

Now we add the two series, term by term, and divide by 2. The resulting expression gives $\cosh x$ as the sum of an infinite series of powers of x:

$$\cosh x = 1 + \frac{x^2}{2!} + \frac{x^4}{4!} + \cdots.$$

Observe that the odd powers of x have dropped away; only the even powers of x are left.

Now suppose that x is not too large, say $|x| < \frac{1}{2}$ (i.e., x lies between $-\frac{1}{2}$ and $\frac{1}{2}$); then quantities such as $x^4/4!$, $x^6/6!$, ... are small in comparison with the first two terms. We conclude that

If $|x|$ is small, then $\cosh x \approx 1 + \frac{1}{2}x^2$.

It follows that for small values of x, the curve $y = \cosh x$ resembles the parabola $y = 1 + \frac{1}{2}x^2$.

For comparison, in Figure 12.9 we have drawn the curves $y = \cosh x$ and $y = 1 + \frac{1}{2}x^2$ on the same pair of axes. The closeness of the two curves for small values of x is plainly visible.

12.6 The logarithmic spiral

The following problem is well known. Let four bugs be seated at the four corners of a square $ABCD$. For convenience we

---

cosh like 'posh' and sinh like 'shine', and these invariably provoke laughter when said aloud in class! Hyperbolic equivalents of the other four trigonometric functions (tan, cot, sec and cosec) also exist; these are the tanh, coth, sech and cosech functions, respectively. The reader should say these names aloud in order to appreciate the difficulties involved.

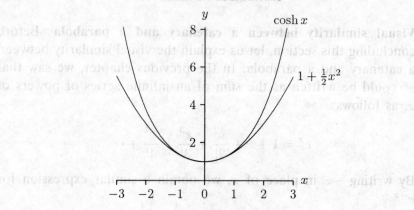

Figure 12.9. *The catenary and parabola*

refer to the four bugs respectively as bug(A), bug(B), bug(C)
and bug(D). At a given instant let each bug start moving; bug(A)
moves directly towards bug(B), bug(B) moves directly towards
bug(C), bug(C) moves directly towards bug(D), and bug(D) moves
directly towards bug(A). They move at the same unvarying speed,
and they continuously change their direction of movement, so that
each one is always moving *directly* towards its "target" bug. By
symmetry, at each instant their positions form the corners of a
square, and the square slowly rotates and shrinks; the bugs move
inwards towards the centre of the square along a spiral path, and
this is where they ultimately meet. Question: *What is the nature of
the path followed by the bugs?* That is, what *kind* of spiral is it?

It is easy to simulate the situation on a computer screen. The
display in Figure 12.10 shows the paths taken by the four bugs.
Observe that the paths are congruent copies of one another.

The spiral seen in Figure 12.10 is a *logarithmic spiral*, and its
equation may be given in polar coordinates as follows (a brief
note on polar coordinates is given at the end of this section):

$$r = ae^{b\theta}.$$

Here a and b are constants:

- a measures the scale of the spiral;

- b measures the rate at which the spiral turns (i.e., how fast
  it spirals outwards); if $b > 0$, the spiral turns outwards in a
  counterclockwise direction; whereas if $b < 0$, then the spiral
  turns outwards in a *clockwise* direction.

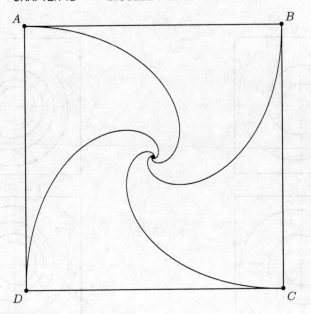

Figure 12.10. *The great bug chase*

Some examples of such spirals, with $a = 1$ and various values of b, are displayed in Figure 12.11.

12.6.1 A brief note on polar coordinates

In polar coordinates, the position of a point P is specified with reference to a *polar axis* ℓ and a *pole* O, which lies on ℓ, by giving its *polar distance*. The distance $r = |OP|$, and its *direction*, i.e., the angle θ from ℓ to the ray OP, is measured in the counterclockwise or positive sense (see Figure 12.12). The polar equation of a curve shows how r and θ are related on the curve. Some examples are given below.

The Archimedean spiral $r = a\theta$, where a is a constant, yields an 'Archimedean spiral' (see Figure 12.13). Its characteristic feature is that as one travels outwards on the spiral, the distance between successive coils stays constant; for this reason, it is also called an 'arithmetic spiral'.

The logarithmic spiral   The defining property of a logarithmic spiral ($r = ae^{\theta}$ where a is a constant) is that the distance between successive coils grows by a fixed ratio after each complete rotation,

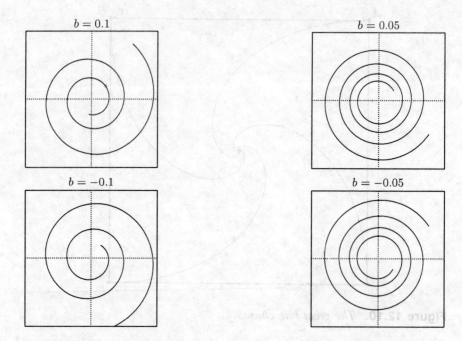

Figure 12.11. *Some logarithmic spirals with varying parameter values*

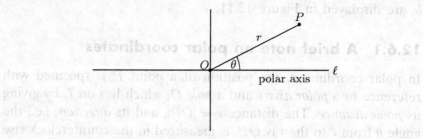

Figure 12.12. *Polar coordinate system*

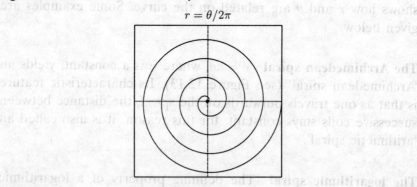

Figure 12.13. *An Archimedean spiral*

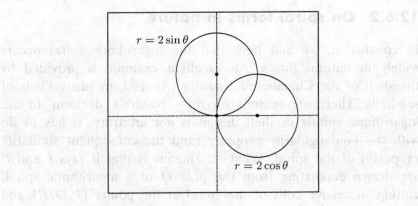

Figure 12.14. *Two circles*

for which reason it is referred to as a 'geometric spiral'. It is also known as an *equiangular spiral*, but this is for another reason. Consider a ray emanating in any direction from the pole O; it intersects the spiral at infinitely many points. Remarkably, the angle between the ray and the curve is the same at all these points. Alternatively phrased, the tangents to the spiral at the various points of intersection are all parallel to one another.

It turns out that the logarithmic spiral is mathematically very rich in properties. It has a near-magical way of reincarnating itself after being subjected to different transformations. To give just one example: the locus of the centres of curvature of a logarithmic spiral (the *evolute* of this spiral) is another logarithmic spiral, congruent to the first one.

Other spirals In contrast, $r = \sqrt{\theta}$ yields 'Fermat's spiral', which grows at a slower rate than the Archimedean spiral (the distance between successive coils of the spiral *decreases* as we move outwards), and $r = \theta^2$ yields a spiral for which the inter-coil distance *increases* as we move outwards (though not as rapidly as in the logarithmic spiral).

Polar equation for a circle Shown in Figure 12.14 are two circles of unit radius, centred at $(0, 1)$ and $(1, 0)$ respectively; their equations are $r = 2\sin\theta$ and $r = 2\cos\theta$.

12.6.2 On spiral forms in nature

In conclusion, we add here that the logarithmic spiral occurs widely in natural forms. An excellent example is provided by the shell of the Chambered Nautilus[3]; indeed, by many kinds of seashells. There are reasons for this—Nature's 'decision' to use logarithmic spirals in shell design is *not* arbitrary; it has to do with the equiangularity property (and the consequent similarity property) of the spiral. What this means is that if rays ℓ and ℓ' are drawn emanating from the pole O of a logarithmic spiral, cutting successive coils of the spiral at the points $\{P, Q, R\}$ and $\{P', Q', R'\}$ respectively, then the shapes $PQQ'P'$ and $QRR'Q'$ are similar to one another (under a suitable expansion of scale). This feature will certainly hold in any natural form that 'chooses' to grow in a spiral shape; for as it grows, its basic proportions are maintained. So if at all a creature adopts a spiral form of growth, the logarithmic spiral would seem to be a natural and inevitable one.

However, not every spiral occurring in Nature is logarithmic. The Archimedean spiral, in which the width between successive coils remains constant, is often seen. Some examples which are particularly appealing are the rolled-up proboscis ("drinking tube") of a butterfly or moth; the rolled-up tail of a chameleon; a snake coiled up in a state of rest or in a defensive pose, resembling a coil of rope; or a rolled-up millipede. Also of great visual appeal is a spider's web, which has the equal-width property though it is made up of segments of straight lines rather than curves.

The spiral arrangement of pollen grains within a sunflower looks extremely appealing, and one may suspect that here too Nature has used a logarithmic model; but recent studies suggest that this is not so. The evidence suggests rather that the relevant spiral in this case has the polar equation $r = a\sqrt{\theta}$ for some constant a.

The three-dimensional version of a spiral is a *helix*, and this form too is often seen, in seashells and the horns of sheep, goats,

---

[3]The Chambered Nautilus lives in the Pacific Ocean at depths of 200 meters or more. It spends its life in a chambered shell, and as it grows it adds new chambers. Each successive chamber is larger than the preceding one, by just the 'right' amount; this accounts for the overall spiral shape. The Nautilus possesses the ability to move up and down within the water by altering its buoyancy; it does this very simply—by altering the fluid–gas ratio in the chambers. It has thus mastered a submarine-like ability.

deer and antelopes; for example, the horns of a Blackbuck or a Kudu. In these examples, there is an outward growth; the helix may also grow only in a perpendicular direction, as in a tendril, or the DNA double helix; here we have no lateral growth—the scale remains constant.

12.7 In conclusion

It should be clear from all that has been said that the exponential and logarithmic functions have great relevance in the natural sciences. Many more examples of the kind studied above may be listed. Here are a few examples.

A damped pendulum   Here we study the amplitude of an oscillating pendulum which is subject to damping; that is, to a frictional force. The amplitude slowly decays over time, and it is found that the decay is exponential in nature.

To be more specific, let the displacement of the pendulum from its mean (central) position be x at time t. In the undamped situation, it is found that x has the following dependence on t:

$$x = a \sin \omega t,$$

where a is the amplitude of motion (its maximum displacement) and ω is the frequency of oscillation. In this situation, the oscillations go on forever, and the amplitude of oscillation remains fixed at a. With damping present, the amplitude slowly decreases over time. We shall assume that the frictional force is proportional to the velocity of the pendulum. Analysis now yields an equation of the form

$$x = e^{-bt} \sin \omega t,$$

where b is a decay constant. The graph of x against time t is shown in Figure 12.15. Observe how the amplitude slowly dies away.

Electric circuits   In this case, we study the growth of current in a circuit with a constant voltage supply. The circuit has a *resistance* R, which 'resists' the current I; an *inductance* L which opposes changes in current via magnetic effects; and a *capacitance* C which stores charge, tends to stabilize the current, and in consequence

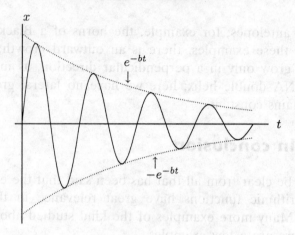

Figure 12.15. *Damped simple harmonic motion*

opposes changes in current. On analysis, it is found that the law of growth of current is exponential in nature. More specifically, if the voltage supplied is V_0 (a constant), then

$$I = \frac{V_0}{R}\left(1 - e^{-Rt/L}\right),$$

where t is time.

The above equation shows that as t increases, I tends towards its steady-state value V_0/R. This is the value predicted by Ohm's law ("current equals voltage divided by resistance"), but it takes time to stabilize at this level.

Variation of atmospheric pressure with height It is a matter of common observation that atmospheric pressure falls with increasing height. We feel it in a plane when it takes off, and again when it starts its descent for landing; our ears are our guide in these matters—when a pressure change occurs, the ears "pop" because of the difference in pressure on the two sides of the eardrum. The same thing happens when the plane hits an "air pocket". (The experience can be painful, especially if one has a cold. Sucking a sweet can help, which is perhaps why the crew distribute candy sweets at the time of takeoff.)

One effect of a fall in atmospheric pressure is a lowering of the boiling point of water. This effect shows in the kitchen: foods get improperly cooked; pulses, in particular, tend to stay uncooked (all mountain climbers will testify to this!). The answer to this problem is to use a pressure cooker. Another effect of a fall in air pressure is increased cloud cover and consequent rainfall; the pressure at the centre of a hurricane or cyclone is extremely low. (During a cyclone, the sea level actually rises—by as much as a meter!) At heights above 4000 m, the pressure is low enough to produce high-altitude sickness, which can be a serious health hazard for mountain climbers.

Atmospheric pressure is defined as the force exerted per unit area by the atmosphere; it is the weight of the entire column of air standing above the unit area in question, and is measured in units of weight/area. It may be measured by measuring the height of a mercury column that the column can support; so it is often given in terms of a height of a column of mercury. It may also be measured by means of a partially evacuated, airtight, corrugated metal box, which responds to changes in atmospheric pressure by contraction or expansion. A needle attached to the box indicates the air pressure.

Atmospheric pressure may be measured in any of the following units: in terms of the height in mm of a column of mercury (Hg); in pounds per square inch (psi); in dynes per square centimeter; in millibars (mb); or in atmospheres (atm). The standard sea-level air pressure may be given in any of the following ways: 760 mm of Hg; 14.70 psi; 1013.25×10^3 dynes/cm^2; 1013.25 mb; 1 atmosphere. Variations in sea-level air pressure are small; the highest ever measured has been 806.45 mm of Hg, in Siberia; and the lowest is 665.1 mm of Hg, in the middle of a cyclone in the South Pacific.

Variations in air pressure at given altitudes are low, so it is possible to talk of a "standard air pressure" corresponding to any particular altitude. It is found that pressure falls roughly by a factor of 10 for every 16000 m of ascent. Near the Earth's surface the pressure drops by about 3.5 mb every 30 m. The table below gives the air pressure at various altitudes.

Height (m)	0	988	2463	5572	10357	16206
Pressure (mb)	1013	900	750	500	250	100

Let P_0 denote the sea-level atmospheric pressure, and P_h the pressure at an altitude of h (measured in m); then we have, *very roughly*,

$$P_h = P_0 \times 10^{-h/16000}.$$

The formula is, we stress, only an approximation. If we use it to compute what P_h should be for various values of h, here is what we obtain:

h (m)	988	2463	5572	10357	16206
P_h via formula (mb)	879	711	454	228	98

The discrepancies between the predicted and the actual values are plainly visible.

The nature of mathematical modeling   Finally, we state the obvious —something which needs to be stated whenever we attempt to study nature in terms of a mathematical model – *No model is perfect*. In particular, no mathematical model is perfect. Inbuilt into the process of modeling is a deliberate blocking of information that is felt to be "not too important". As Alan Turing once said, in describing an attempt to model the phenomenon of morphogenesis,

> This model will be a simplification and an idealization, and consequently a falsification. It is hoped that the features retained for discussion are those of the greatest importance in the present state of knowledge.

And that surely provides a fitting note upon which to end this chapter.

Chapter 13

An Areal View

In this chapter, we shall present another approach to the logarithmic function—axiomatic and geometric in spirit.

13.1 The function L(t)

Consider the graph whose equation is $y = 1/x$ for $x > 0$ (see Figure 13.1). For $t > 0$, consider the region enclosed by the curve, the x-axis, and the lines $x = 1$, $x = t$. We call the region $\mathcal{R}(t)$.

Let $L(t)$ denote the area of this region, which we enclose by travelling from $(1, 0)$ to $(t, 0)$ along the x-axis; up along the line $x = t$ to $(t, 1/t)$; along the curve to $(1, 1)$; and then down the line $x = 1$ to $(1, 0)$. Obviously, $L(1) = 0$.

By convention, when we traverse the boundary of a region in a 'positive' or counter-clockwise direction, its area is taken to be

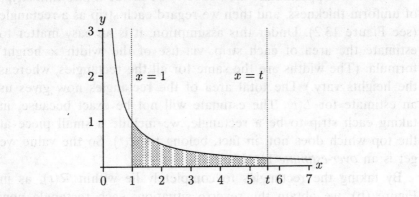

Figure 13.1. *Graph of* $y = 1/x$

153

positive, and when we traverse the boundary in a 'negative' or clockwise direction, its area is taken to be negative. This is the normal convention followed in coordinate geometry; the area thus obtained is referred to as its *oriented area*.

An immediate implication of the convention, as applied to the function $L(t)$, is the following: for $t > 0$,

$$L(t) \text{ is } \begin{cases} \text{negative} & \text{if } t < 1; \\ \text{positive} & \text{if } t > 1. \end{cases}$$

The reason for this is that if $t < 1$, then the path taken has a clockwise direction; whereas if $t > 1$, then the path has a counter-clockwise direction. The cross-over point, where the values change from negative to positive, is $t = 1$.

Having introduced the function L, we will naturally want to see what sort of function it is; i.e., what values it takes, and what is the manner of its variation. We start by computing $L(t)$ for different values of t.

13.2 Approximate values

An immediate difficulty faced in computing $L(t)$ is that the region $\mathcal{R}(t)$ is *curved but not circular*; so none of the usual formulas of elementary geometry will yield its area. This being the case, we shall attempt to *estimate* $L(t)$ rather than compute it exactly.

To find the approximate area of $\mathcal{R}(t)$ we divide it into thin strips of uniform thickness, and then we regard each strip as a rectangle (see Figure 13.2). Under this assumption, it is an easy matter to estimate the area of each strip, via use of the 'width × height' formula. (The widths are the same for all the rectangles, whereas the heights vary.) The total area of the rectangles now gives us an estimate for $L(t)$. The estimate will not be exact because, in taking each strip to be a rectangle, we include a small piece at the top which does not, in fact, belong to $\mathcal{R}(t)$. So the value we get is an *over-estimate* for $L(t)$.

By taking the rectangles to completely lie within $\mathcal{R}(t)$, as in Figure (b), we obtain the reverse situation: each rectangle now *excludes* a small portion of the region, so the sum of the areas of these rectangles provides an *under-estimate* for $L(t)$. Obviously, $L(t)$ lies between these two estimates.

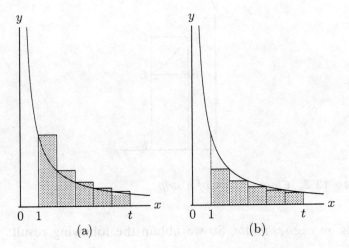

Figure 13.2. *Estimating $L(t)$ from both ends*

By this means, we have succeeded in enclosing the true (un-known) value of $L(t)$ within an interval; now we must find ways of narrowing the interval. This, fortunately, is easy to do: we simply reduce the strip width. We shall show below that the desired effect does indeed take place.

Choose the width w of each region to be $1/n$ where n is a large positive integer; then the vertical lines that divide the region are $x = 1 + 1/n$, $x = 1 + 2/n$, $x = 1 + 3/n$, $\ldots$, and the kth line is $x = 1 + k/n$. The close-up in Figure 13.3 shows the kth strip, bounded by the line $x = 1 + (k-1)/n$ on the left (the line ABC), and by $x = 1 + k/n$ on the right (the line FED).

The coordinates of A, B, C, D, E and F are computed from the equation of the curve. We get

$$AF = \frac{1}{n}, \quad AB = EF = \frac{n}{n+k}, \quad AC = FD = \frac{n}{n+k-1},$$

so the area of the rectangle $ACDF$ is $AF \times AC$ or $1/(n+k-1)$, and the area of the rectangle $ABEF$ is $AF \times AB$ or $1/(n+k)$.

Since the total width of $\mathcal{R}(t)$ is $t-1$ and each strip has a width of $1/n$, the total number of strips is $(t-1) \div \text{width} = (t-1)n$. (We shall assume for simplicity that $(t-1)n$ is an integer.) So our procedure for estimating $L(t)$ is simply to find the sum of the reciprocals of the $(t-1)n$ integers n, $n+1$, $n+2$, $\ldots$, $tn-1$. This

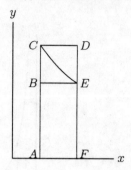

Figure 13.3. *Close-up view of a strip*

yields an *over-estimate*. So we obtain the following result:

$$L(t) < \frac{1}{n} + \frac{1}{n+1} + \frac{1}{n+2} + \cdots + \frac{1}{nt-1}.$$

The quantity on the right side will be denoted by $L^+(t;n)$. (The plus sign indicates that it is an upper estimate, and the 'n' indicates the dependence of the estimate on the number of strips.)

For the lower estimate, we consider the rectangle $ABEF$, whose area is $AF \times AB = 1/(n+k)$. Summing up these areas for $k = 1$, 2, 3, ..., $(t-1)n$, we get the desired under estimate, which we denote by $L^-(t;n)$:

$$L(t) > \frac{1}{n+1} + \frac{1}{n+2} + \frac{1}{n+3} + \cdots + \frac{1}{nt}.$$

The minus sign indicates that $L^-(t;n)$ is an under estimate, and the 'n' indicates the dependence of the estimate on n.

Observe that the quantities $L^-(t;n)$ and $L^+(t;n)$ are both sums of $(t-1)n$ fractions, and that they are *nearly* the same. Indeed, they differ only in the first and last terms, and we have

$$L^+(t;n) - L^-(t;n) = \left(\frac{1}{n} + \frac{1}{nt-1} \right) - \left(\frac{1}{n+1} + \frac{1}{nt} \right).$$

After doing the necessary simplifications, we find that when n is large the quantity on the right is roughly equal to

$$\frac{1}{n^2} \left(1 + \frac{1}{t^2} \right).$$

Since t is a fixed quantity, we see that the gap between the lower and upper estimates vary roughly as $1/n^2$. This shows that the gap rapidly shrinks as n grows larger. By choosing a suitably large value for n, we may achieve any desired level of accuracy.

We now have the necessary prescription for estimating $L(t)$, and we must put it into action.

Estimating L(2)

Since $L(2)$ lies between $L^-(2;n)$ and $L^+(2;n)$, where

$$L^-(2;n) = \frac{1}{n+1} + \frac{1}{n+2} + \cdots + \frac{1}{2n-1} + \frac{1}{2n},$$

$$L^+(2;n) = \frac{1}{n} + \frac{1}{n+1} + \cdots + \frac{1}{2n-2} + \frac{1}{2n-1},$$

we compute the table displayed below.

n	100	1000	10000
$L^-(2;n)$	0.6907	0.6929	0.6931
$L^+(2;n)$	0.6957	0.6934	0.6932

We conclude that $L(2) \approx 0.69315$.

Estimating L(3)

Since $L(3)$ lies between $L^-(3;n)$ and $L^+(3;n)$, where

$$L^-(3;n) = \frac{1}{n+1} + \frac{1}{n+2} + \cdots + \frac{1}{3n-1} + \frac{1}{3n},$$

$$L^+(3;n) = \frac{1}{n} + \frac{1}{n+1} + \cdots + \frac{1}{3n-2} + \frac{1}{3n-1},$$

we compute the table displayed below.

n	100	1000	10000
$L^-(3;n)$	1.095	1.0983	1.09858
$L^+(3;n)$	1.102	1.0989	1.09864

We conclude that $L(3) \approx 1.0986$.

Estimating L(4)

Since $L(4)$ lies between $L^-(4;n)$ and $L^+(4;n)$, where

$$L^-(4;n) = \frac{1}{n+1} + \frac{1}{n+2} + \cdots + \frac{1}{4n-1} + \frac{1}{4n},$$

$$L^+(4;n) = \frac{1}{n} + \frac{1}{n+1} + \cdots + \frac{1}{4n-2} + \frac{1}{4n-1},$$

we compute the table displayed below.

n	100	1000	10000
$L^-(4;n)$	1.383	1.386	1.38626
$L^+(4;n)$	1.390	1.387	1.38633

We conclude that $L(4) \approx 1.3863$. Remarkably, this is exactly twice the value we had obtained for $L(2)$!

Estimating L(5)

Since $L(5)$ lies between $L^-(5;n)$ and $L^+(5;n)$, where

$$L^-(5;n) = \frac{1}{n+1} + \frac{1}{n+2} + \cdots + \frac{1}{5n-1} + \frac{1}{5n},$$

$$L^+(5;n) = \frac{1}{n} + \frac{1}{n+1} + \cdots + \frac{1}{5n-2} + \frac{1}{5n-1},$$

we compute the table displayed below.

n	100	1000	10000
$L^-(5;n)$	1.605	1.6090	1.6094
$L^+(5;n)$	1.613	1.6098	1.6095

We conclude that $L(5) \approx 1.60945$.

Estimating L(6), L(7) and L(8)

Taking $t = 6$, 7 and 8 in turn, and $n = 10000$, we arrive at the following data:

- $\underline{t = 6}$
  lower estimate $= 1.7917$, upper estimate $= 1.7918$;

- $\underline{t = 7}$
  lower estimate $= 1.9459$, upper estimate $= 1.9459$;

- $\underline{t = 8}$
  lower estimate $= 2.0793$, upper estimate $= 2.0795$.

We conclude that $L(6) \approx 1.79175$, $L(7) \approx 1.9459$, and $L(8) \approx 2.0794$.

   Remarkably, we find that $L(6)$ is very nearly equal to $L(2)+L(3)$; for

$$L(2) + L(3) \approx 0.69315 + 1.0986 = 1.79175;$$

and $L(8)$ is very nearly equal to $3 \times L(2)$; for

$$3 \times L(2) \approx 3 \times 0.69315 = 2.07945.$$

Is this coincidence or (as James Bond might have said) 'happen-stance'?

Observing that $L(4)$ is 2 times $L(2)$ and that $L(8)$ is 3 times $L(2)$, we may wonder whether $L(16)$ is 4 times $L(2)$. So we use the computer once again, this time to compute $L(16)$. Here are the results, using $n = 10000$:

lower estimate $= 2.7725$,   upper estimate $= 2.7726$.

Our question is answered: we find that $L(16)$ is indeed equal (very nearly) to 4 times $L(2)$; for $4 \times L(2) = 4 \times 0.69315 = 2.7726$.

Clearly, there is some underlying law at work here.

13.3   Properties of L(t)

Based on the preceding comments, we may surmise that

- $L(ab) = L(a) + L(b)$ for any positive numbers a, b;

- $L(t^k) = kL(t)$ for any positive number t and any real number k.

These laws look familiar—why, they are just the laws that hold for logarithms! Good gracious! Are we, then, dealing with a logarithmic function in disguise? We are!—and we shall now prove it.

It suffices to prove the second property, for the first follows from the second. (Indeed, each property follows from the other.) For, *given* that the second statement holds, write $a = 2^\alpha$ and $b = 2^\beta$ for appropriate numbers α and β (indeed, $\alpha = \log_2 a$ and $\beta = \log_2 b$). Then, we have

$$ab = 2^\alpha \cdot 2^\beta = 2^{\alpha+\beta},$$
$$\therefore \quad L(ab) = L(2^{\alpha+\beta}) = (\alpha + \beta) L(2).$$

Also, $L(a) = L(2^\alpha) = \alpha L(2)$ and $L(b) = L(2^\beta) = \beta L(2)$, $\therefore L(a) + L(b) = (\alpha + \beta) L(2) = L(ab)$. So the first property follows from the second.

COMMENT. The '2' used above, as the base, has no significance; we could have used 3 or 10, or indeed any positive number not equal to 1.

There are several ways of proving the second property; some are geometric in nature, some are algebraic, some use calculus; but all are more or less equivalent—what they do 'at bottom' is more or less the same. We shall initially opt for an algebraic proof.

Let us start by explaining why $L(4) = 2L(2)$. We have seen that for large values of n, using the formula for the lower estimate in each case, we have

$$L(2) \approx \frac{1}{n+1} + \frac{1}{n+2} + \cdots + \frac{1}{2n}$$

$$L(4) \approx \left(\frac{1}{n+1} + \frac{1}{n+2} + \cdots + \frac{1}{2n} \right)$$

$$+ \left(\frac{1}{2n+1} + \frac{1}{2n+2} + \cdots + \frac{1}{4n-1} + \frac{1}{4n} \right).$$

In the equation for $L(4)$, the first parenthesis has n terms while the second parenthesis has $2n$ terms ($3n$ terms in all). Now, observe that

$$\left(\frac{1}{2n+1} + \frac{1}{2n+2} \right) - \frac{1}{n+1} = \frac{1}{(2n+1)(2n+2)} < \frac{1}{4n^2},$$

$$\left(\frac{1}{2n+3} + \frac{1}{2n+4} \right) - \frac{1}{n+2} = \frac{1}{(2n+3)(2n+4)} < \frac{1}{4n^2},$$

$$\left(\frac{1}{2n+5} + \frac{1}{2n+6} \right) - \frac{1}{n+3} = \frac{1}{(2n+5)(2n+6)} < \frac{1}{4n^2},$$

and so on. We see that each time the sum of the two terms we pick from $L(4)$ exceeds the corresponding term from $L(2)$ by a small margin. The difference at each stage is never more than $1/4n^2$, and since there are n quantities of this sort, the total error is no more than $1/4n$. So the following holds: *For each n, the difference between $L^-(4; n)$ and $2 \times L^-(2; n)$ is no more than $1/4n$.* As n increases, this difference shrinks to 0, and we conclude that $L(4) = 2L(2)$.

We may prove in the same way that $L(9) = 2L(3)$. For large n, we have

$$L(3) \approx \frac{1}{n+1} + \frac{1}{n+2} + \cdots + \frac{1}{3n},$$

and

$$L(9) \approx \left(\frac{1}{n+1} + \frac{1}{n+2} + \cdots + \frac{1}{3n} \right)$$
$$+ \left(\frac{1}{3n+1} + \frac{1}{3n+2} + \frac{1}{3n+3} + \cdots + \frac{1}{9n} \right).$$

This time, we group the terms in $L(9)$ into threes. We find that

$$\frac{1}{3n+1} + \frac{1}{3n+2} + \frac{1}{3n+3}$$

is very nearly equal to $1/(n+1)$. Indeed, the two quantities differ by

$$\frac{9n+5}{(3n+1)(3n+2)(3n+3)},$$

which is roughly equal to (and does not exceed) $1/3n^2$. So the argument goes through exactly as it did earlier, and the conclusion that $L(9) = 2L(3)$ follows.

The same idea works in proving that $L(t^2) = 2L(t)$ for any positive number t. We shall leave it to the reader to prove in the same way that $L(t^3) = 3L(t)$, $L(t^4) = 4L(t)$, and so on, for any $t > 0$. The general argument is identical in all these cases.

We now take as proven the statement that $L(t^k) = kL(t)$ for all $t > 0$ and all k, and therefore also the statement that $L(ab) = L(a) + L(b)$ for all positive numbers a and b.

★ ★ ★

We shall now show *geometrically* the property that for any two numbers $a, b > 0$ we have $L(ab) = L(a) + L(b)$. The argument is a very pretty one; it directly uses the definition of $L(t)$ as the area enclosed by the curve $y = 1/x$, the x-axis, and the lines $x = 1$ and $x = t$.

Clearly, proving that $L(ab) = L(a) + L(b)$ is identical to proving that $L(ab) - L(b) = L(a)$. Now the quantity $L(ab) - L(b)$ is the area of the region enclosed by the curve, the x-axis, and the lines $x = b$ and $x = ab$. So what we need to show is the following: *The area enclosed by the curve, the x-axis and the lines $x = 1$, $x = a$ equals the area enclosed by the curve, the x-axis and the lines $x = b$, $x = ab$.* That is, we must show that the regions marked I, II in Figure 13.4 have equal area.

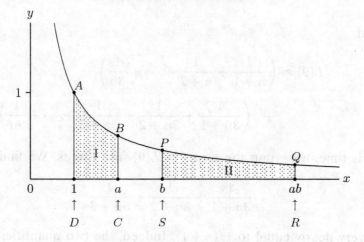

Figure 13.4. *Regions I and II have equal area*

We first label the vertices of I and II as follows:

$$A = (1,1), \qquad B = (a, 1/a), \qquad C = (a, 0), \qquad D = (1, 0);$$
$$P = (b, 1/b), \quad Q = (ab, 1/ab), \quad R = (ab, 0), \quad S = (b, 0).$$

We now use ideas from *transformation geometry* to do the needful. Consider a contraction by a factor of b applied along the x-axis, followed by an expansion by a factor of b applied along the y-axis. That is, first contract in the x-direction by a factor of b, then expand in the y-direction by a factor of b. The net result of the two transformations is that shapes get distorted (they get thinner but taller); however, *their areas remain unchanged*.

Symbolically, we may describe the two operations by the equations $(x, y) \mapsto (x/b, y)$ and $(x, y) \mapsto (x, by)$, respectively. Acting together they produce the following effect:

$$(x, y) \mapsto \left(\frac{x}{b}, y \right) \mapsto \left(\frac{x}{b}, by \right).$$

So the net result is: $(x, y) \mapsto (x/b, by)$. For convenience, we refer to the x-axis contraction by f, and to the y-axis expansion by g.

For any point X, denote by X' the image of X under f, and by X'' the image of X' under g. That is, $f(X) = X'$ and $g(X') = X''$, and $X \mapsto X' \mapsto X''$ under $g \circ f$. Let us now see what happens to the region $PQRS$ under $g \circ f$.

Under f, we get $PQRS \mapsto P'Q'R'S'$, where

$$P' = \left(1, \frac{1}{b} \right), \qquad Q' = \left(a, \frac{1}{ab} \right), \qquad R' = (a, 0), \qquad S' = (1, 0).$$

Observe that only the x-coordinate undergoes a change, and that $R' = C$ and $S' = D$; i.e, $R \mapsto C$, $S \mapsto D$. So the image of region II is a region II$'$ with the same base as region I. It sits inside region I, and its left border falls on the left border of I, while its right border falls on the right border of I.

Under g, we get $P'Q'R'S' \mapsto P''Q''R''S''$, where

$$P'' = (1,1), \quad Q'' = \left(a, \frac{1}{a}\right), \quad R'' = (a,0), \quad S'' = (1,0).$$

Surprise!—we find that $P'' = A$, $Q'' = B$, $R'' = C$ and $S'' = D$. It is not hard to see that the curved upper border of II$'$ is mapped point by point onto the curved upper border of I by $g \circ f$. So II$''$ = I, and the net result is: I $\mapsto$ II.

Therefore, the net effect of the two operations is to map I onto II. We had remarked earlier that under the combined effect of the two operations, the areas remain unchanged. Therefore, the areas of regions I and II are equal! The required result has been proved.

13.4 For which t is L(t) equal to 1?

It is obvious geometrically that $L(t)$ increases steadily with t; for as t increases, the region $\mathcal{R}(t)$ encloses more and more. The calculations done earlier gave the following values (all accurate to 4 decimal places): $L(1) = 0$, $L(2) = 0.69315$, $L(3) = 1.0986$, $L(4) = 1.3863$, $L(5) = 1.60944$, ... The increasing profile is evident in these figures. Observe that $L(2) < 1$ and $L(3) > 1$. We may naturally wonder where (if at all) it happens that $L(t) = 1$. Presumably, $L(t) = 1$ for some value of t lying between 2 and 3.

For starters, we try $t = 2.5$; here is the result. Choosing $n = 10000$ we get: lower estimate = 0.91626, upper estimate = 0.91632. So $L(2.5) \approx 0.9163 < 1$, and we conclude that the desired t lies between 2.5 and 3.

Let us then try $t = 2.75$. Choosing $n = 10000$ as earlier, we get: lower estimate = 1.01157, upper estimate = 1.01163. So $L(2.75) \approx 1.0116$; we have crossed over! The desired value therefore lies between 2.5 and 2.75, and presumably it lies closer to 2.75 than to 2.5.

Further experimentation may be done along these lines, but we shall opt for another approach, making use of our all-important

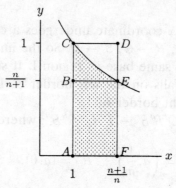

Figure 13.5. *Close-up view of a strip*

discovery that $L(t^k) = kL(t)$.

★ ★ ★

Let n be a large positive integer. What can be said about the value of $L(1 + 1/n)$? It refers to the area enclosed in the narrow strip bounded by the lines $x = 1$ and $x = 1 + 1/n$, and by the curve $y = 1/x$ and the x-axis. A close-up view is shown in Figure 13.5. We see that

$$AF = \frac{1}{n}, \quad EF = AB = \frac{n}{n+1}, \quad AC = FD = 1.$$

From this, it follows that the area of the rectangle $ACDF$ is $AF \times AC = 1/n$, and the area of rectangle $ABEF$ is $AF \times AB = 1/(n+1)$. Therefore,

$$\frac{1}{n+1} < L\left(1 + \frac{1}{n}\right) < \frac{1}{n}.$$

Multiplying through by n, we conclude that $nL(1+1/n)$ lies between $n/(n+1)$ and 1.

Next, from the property that $L(t^k) = kL(t)$, we deduce that $nL(1+1/n)$ equals $L(a_n)$, where $a_n = (1+1/n)^n$. We have, therefore,

$$\frac{n}{n+1} < L(a_n) < 1.$$

Observe that in this double inequality the quantities at the ends, namely $n/(n+1)$ and 1, are very close to one another. As n gets larger, the ends get closer to one another; the interval between the ends steadily shrinks. Such behaviour is often seen in advanced mathematics, and mathematicians invoke the so-called "sandwich principle" to conclude that in the limit the interval has zero width—it is 'squeezed out' completely. It follows from this that the L-values of the quantities

$$\left(1+\frac{1}{2}\right)^2, \quad \left(1+\frac{1}{3}\right)^3, \quad \ldots, \left(1+\frac{1}{100}\right)^{100}, \quad \ldots$$

get ever closer to 1, and the L-value of the limiting value of a_n actually equals 1.

Now, we have seen the sequence $a_1, a_2, \ldots, a_n, \ldots$ earlier, in Section 10.1, where we had *defined* e as the limiting value of the quantities $(1 + 1/n)^n$. So we reach the following remarkable conclusion: $L(e) = 1$.

13.5 Identification of L(t)

We are now in a position to completely identify L.

Applying the law $L(t^k) = kL(t)$ to the case when $t = e$, we deduce that

$$L(e^2) = 2, \quad L(e^3) = 3, \quad L(e^4) = 4, \quad \ldots,$$

and, in general, $L(e^k) = k$ for any k.

Now, how is k related to the quantity e^k? That is, given the quantity e^k, how can k be recovered from it? Clearly, by taking natural logarithms; i.e., by the equation $k = \ln e^k$. Thus, at long last we reach our desired conclusion:

$$L(t) = \ln t \quad \text{for all } t > 0.$$

So L is a logarithmic function, just as we had suspected.

EXERCISES

13.5.1 Show how, from the property that $L(ab) = L(a) + L(b)$ for any two positive numbers a, b, there follows the property that $L(t^k) = kL(t)$ for any $t > 0$ and any k.

13.5.2 Explain why, after a contraction along the x-axis by a factor of b and an expansion along the y-axis by a factor of b, the areas remain unchanged (though shape distortion does take place).

Appendix A

The Irrationality of e

In what follows, we shall prove two results about e. From the point of view of applications in science and technology (the "practical point of view"), these results have no particular significance[1], but from an arithmetical point of view (particularly that of number theory), they carry great interest and significance.

Remarks on Irrationality   We shall first briefly discuss irrationality. A number x is *rational* if it can be written in the form $x = a/b$, where a and b are non-zero integers. For example, $1.23 = 123/100$ is rational, and so is $3.567 = 3567/1000$. It is known that an infinite repeating decimal fraction represents a rational number. For example,

$$0.12\,12\,12\,12 \ldots = \frac{12}{99} = \frac{4}{33}.$$

A number is said to be *irrational* if it is not rational. Observe that this is a "negative definition"; it is not clear, using only the definition, that there exist numbers that are irrational, or how a given number may be shown to be irrational. The very existence of such numbers is far from obvious.

It was the Pythagorean school of geometry that first discovered the existence of irrational numbers; they showed that $\sqrt{2}$ is irrational. Other examples of irrational numbers are $\sqrt{3}$ and $\sqrt{5}$. The well-known number π is irrational (hence the statement that

---

[1] As far as physical measurements are concerned, the concept of irrationality has no relevance whatsoever.

$\pi = 22/7$ is only an approximation), but proving this is very hard. The proof that $\sqrt{2}$ is irrational is given below.

Irrationality of the square root of 2   This well-known and beautiful proof uses an idea pioneered by the Greeks—that of "proof by contradiction". It *supposes* that $\sqrt{2}$ is rational, then proceeds to show that the supposition contradicts itself; that is, it "crashes under its own weight". This shows that the supposition must be false (for self-contradiction is not permitted in mathematics) and, therefore, that $\sqrt{2}$ is indeed irrational. Here are the details of the argument.

Suppose that $\sqrt{2} = a/b$, where a, b are positive integers with no factors in common. From the relation we get $a^2 = 2b^2$, $\therefore$ a^2 is an even number. Since the square of an odd number is odd, it follows that a is odd; $\therefore$ $a = 2c$, where c is some integer. Substitution now yields $2b^2 = 4c^2$, or $b^2 = 2c^2$; $\therefore$ b^2 is an even number, so b too must be even. So a *and* b *are both even*, which means that they have a factor in common, contradicting what we said at the start about their having no factors in common. The self-contradiction shows that the initial supposition must be false and consequently that $\sqrt{2}$ is irrational.

Irrationality of e   We now show that it is not possible to find non-zero integers a and b such that $e = a/b$.

We shall suppose that e is rational; i.e., that there do exist positive integers a and b such that $e = a/b$. Then we shall show that the assumption is self-contradictory.

Suppose that coprime positive integers a and b exist such that $e = a/b$. Since e is not an integer (we know that $2 < e < 3$), we have $b > 1$ ($\therefore$ $b \geq 2$). Now, recall that e is given by the sum of the following infinite series:

$$e = 1 + \frac{1}{1!} + \frac{1}{2!} + \frac{1}{3!} + \frac{1}{4!} + \cdots .$$

Multiplying both sides of this equation by $b!$ we find that $b!e = A + B$, where

$$A = b! + \frac{b!}{1!} + \frac{b!}{2!} + \frac{b!}{3!} + \cdots + \frac{b!}{b!}$$

$$B = \frac{b!}{(b+1)!} + \frac{b!}{(b+2)!} + \frac{b!}{(b+3)!} + \cdots .$$

Observe that A is a sum of integers and so is an integer; $b!e$ too is an integer (since $b!e = (b-1)!a$), and therefore so is $B = A - b!e$ (indeed B is a *positive* integer). Now observe that

$$\frac{(b+1)!}{b!} = b+1 > b, \qquad \frac{(b+2)!}{b!} = (b+1)(b+2) > b^2,$$

$$\frac{(b+3)!}{b!} = (b+1)(b+2)(b+3) > b^3, \qquad \ldots$$

From this, it follows that $B < \dfrac{1}{b} + \dfrac{1}{b^2} + \dfrac{1}{b^3} + \cdots$.

Now recall that $t + t^2 + t^3 + \cdots = t/(1-t)$ for $t > 1$. Using this, we get

$$B < \frac{1/b}{1 - 1/b} = \frac{1}{b-1} \leq 1 \quad \text{(since } b \geq 2), \qquad \therefore \quad 0 < B < 1.$$

Earlier we had noted that B must be a positive integer, and now we find that $0 < B < 1$; a contradiction! The desired end has been achieved; we have shown that the supposition that e is rational is self-contradictory. It follows from this argument that e is irrational.

Stronger results The second result says more, but it is harder to prove. We shall show that *it is not possible to find integers a, b, c, not all zero, such that $ae^2 + be + c = 0$.* Note that this statement implies that e is irrational (simply take a to be 0); hence the adjective "stronger". The proof is from [Bea] (see the Bibliography).

As earlier, we shall start with the infinite series for e and $1/e$:

$$e = 1 + \frac{1}{1!} + \frac{1}{2!} + \frac{1}{3!} + \frac{1}{4!} + \cdots,$$

$$\frac{1}{e} = \frac{1}{2!} - \frac{1}{3!} + \frac{1}{4!} - \frac{1}{5!} + \cdots.$$

Choose any *odd* positive integer $n > 2$ and multiply both series by $n!$; we get

$$n!e = A + \frac{1}{n+1} + \frac{1}{(n+1)(n+2)} + \cdots,$$

$$\frac{n!}{e} = B + \frac{1}{n+1} - \frac{1}{(n+1)(n+2)} + \cdots,$$

where A and B are integers. Write $u = n!e - A$ and $v = n!/e - B$; then

$$u = \frac{1}{n+1} + \frac{1}{(n+1)(n+2)} + \cdots,$$

$$v = \frac{1}{n+1} - \frac{1}{(n+1)(n+2)} + \cdots.$$

We now have $1/(n+1) < u < 1/n$; the first inequality is obvious, and for the second:

$$u < \frac{1}{n+1} + \frac{1}{(n+1)^2} + \frac{1}{(n+1)^3} + \cdots = \frac{1/(n+1)}{1 - 1/(n+1)} = \frac{1}{n}.$$

This establishes the upper bound for u. Next we have

$$\frac{1}{n+2} < v < \frac{1}{n+1}.$$

The inequality $v > 1/(n+2)$ follows from the fact the first two terms of v yield $1/(n+2)$ and the rest of the series yields a positive number (to see why, group the terms in pairs; in each pair the first term exceeds the second term). The inequality $v < 1/(n+1)$ follows in exactly the same way.

So u lies between $1/(n+1)$ and $1/n$, and v lies between $1/(n+2)$ and $1/(n+1)$. It follows that we can write u and v in the following form:

$$u = \frac{1}{n+\alpha_n}, \quad v = \frac{1}{n+1+\beta_n},$$

where α_n and β_n are numbers lying strictly between 0 and 1 (the subscript shows their dependence on n).

Now we suppose that some integers a, b, c, not all zero, exist such that $ae^2 + be + c = 0$. Multiplying this equation by the factor $n!/e$ (where n is the odd positive integer used above), we get

$$a(n!e) + bn! + c\left(\frac{n!}{e}\right) = 0.$$

Since $n!e = A + 1/(n+\alpha_n)$ and $n!/e = B + 1/(n+1+\beta_n)$, we get, by substitution,

$$aA + \frac{a}{n+\alpha_n} + bn! + cB + \frac{c}{n+1+\beta_n} = 0.$$

Since $bn! + cB$ is an integer, it follows that

$$\frac{a}{n + \alpha_n} + \frac{c}{n + 1 + \beta_n} = \text{some integer, say } I_n.$$

Here 'I' stands for integer, and the subscript indicates that the actual integer value taken depends on n.

Now a and c are fixed, and α_n and β_n always lie between 0 and 1; therefore, when n is very large, I_n will be very small ($I_n \approx 0$). This means that when n is very large we must have $I_n = 0$; for the only "very small" integer is 0. Therefore, for all sufficiently large values of n we must have

$$\frac{a}{n + \alpha_n} + \frac{c}{n + 1 + \beta_n} = 0, \quad \therefore \quad \frac{a}{c} = -\frac{n + \alpha_n}{n + 1 + \beta_n}.$$

Since the quantity $(n + \alpha_n)/(n + 1 + \beta_n)$ tends to 1 as n gets indefinitely large, it follows that $a/c = -1$, $\therefore$ $a = -c$. Therefore, we get

$$\frac{a}{n + \alpha_n} = \frac{a}{n + 1 + \beta_n}.$$

This means that $\alpha_n = 1 + \beta_n$, and this is absurd, because α_n and β_n both lie between 0 and 1.

The desired self-contradiction has been found, and the result now follows.

★ ★ ★

In fact, still more is true; it may be shown that *there is no polynomial equation of the form*

$$ax^n + bx^{n-1} + cx^{n-2} + \cdots + k = 0,$$

where $a, b, c, \ldots, k$ are integers, not all zero, and n is a positive integer, for which e is a root. A number which is a root of a polynomial equation with integral coefficients is known as an *algebraic number*, and a number which is not a root of any such equation is described (rather exotically!) as a *transcendental number*. So the statement made above may be stated compactly as: *e is not an algebraic number*, or: *e is transcendental*. However, this statement is extremely hard to prove.

It may be shown that π too is transcendental. Curiously, the proof is considerably more difficult than the corresponding proof for e.

Appendix B

An Infinite Series for e

In Chapter 10, we claimed that

$$e = 1 + \frac{1}{1!} + \frac{1}{2!} + \frac{1}{3!} + \frac{1}{4!} + \cdots,$$

and, more generally, that

$$e^x = 1 + \frac{x}{1!} + \frac{x^2}{2!} + \frac{x^3}{3!} + \frac{x^4}{4!} + \cdots,$$

for all real numbers x. We now show why these statements are true. To do so, we shall make use of the *binomial theorem* (enunciated below).

The binomial theorem Examining, for various positive integral values of n, the expansions of $(1+x)^n$ in powers of x, we quickly perceive a pattern:

$$(1+x)^1 = 1 + x,$$
$$(1+x)^2 = 1 + 2x + x^2,$$
$$(1+x)^3 = 1 + 3x + 3x^2 + x^3,$$
$$(1+x)^5 = 1 + 5x + 10x^2 + 10x^3 + 5x^4 + x^5,$$
$$(1+x)^6 = 1 + 6x + 15x^2 + 20x^3 + 15x^4 + 6x^5 + x^6,$$

and so on. We observe that in each case,

- the coefficient of x is n;

- the coefficient of x^2 is $\dfrac{n(n-1)}{2!}$;

171

- the coefficient of x^3 is $\dfrac{n(n-1)(n-2)}{3!}$;

and so on. (The quantities $2!, 3!, 4!, \ldots$ are the factorial numbers; we have encountered them earlier, in Chapter 11.)

Long ago, Isaac Newton showed that, in general, the following formula holds for all positive integers n:

$$(1+x)^n = 1 + nx + \frac{n(n-1)}{2!}x^2 + \frac{n(n-1)(n-2)}{3!}x^3$$

$$+ \frac{n(n-1)(n-2)(n-3)}{4!}x^4 + \cdots.$$

Here the coefficient of x^r is the product of r fractions, with numerators $n, n-1, n-2, \ldots, n-r+1$ and denominators $1, 2, 3, \ldots, r$ (respectively):

$$\frac{n}{1} \times \frac{n-1}{2} \times \frac{n-2}{3} \times \cdots \times \frac{n-r+1}{r}.$$

This quantity is usually denoted by the symbol $\binom{n}{r}$. Observe that since n is a positive integer, we have

$$\binom{n}{n+1} = 0, \quad \binom{n}{n+2} = 0, \quad \binom{n}{n+3} = 0, \quad \ldots$$

These relations hold by virtue of the zero that appears in the numerator of the above formula when $r = n+1$, $n+2$, $n+3$, $\ldots$; indeed, $\binom{n}{r} = 0$ for all $r > n$. Therefore, when n is a positive integer, the expansion for $(1+x)^n$ is finite—it stops with the term x^n.

Newton's claim is not hard to prove; all one needs is the proof technique known as *mathematical induction*. What is really surprising, however, is that the theorem holds even when n is not zero or a positive integer. The only condition that must be added is: "provided that $-1 < x < 1$". Let us see why such a condition must be added.

First observe that if n is not a positive integer (i.e., it is negative or non-integral), then for no value of r does the coefficient of x^r become 0; that is,

$$\frac{n}{1} \times \frac{n-1}{2} \times \cdots \times \frac{n-r+1}{r} \neq 0$$

for all non-negative integers r. This means that the series goes on forever!

EXAMPLE. Let us see what happens when $n = 1/2$; we get

$$(1+x)^{1/2} = 1 + \frac{1/2}{1!}x + \frac{1/2 \times (-1/2)}{2!}x^2$$

$$+ \frac{(1/2) \times (-1/2) \times (-3/2)}{3!}x^3 +$$

$$+ \frac{(1/2) \times (-1/2) \times (-3/2) \times (-5/2)}{4!}x^4 + \cdots$$

$$= 1 + \frac{x}{2} - \frac{x^2}{8} + \frac{x^3}{16} - \frac{5x^4}{128} + \cdots.$$

The series on the right is an infinite series, and if $-1 < x < 1$ then its sum gives the value of $\sqrt{1+x}$. Some evidence for this will be seen if we do a few computations, say with $x = 0.1$. We have: $\sqrt{1+0.1} = \sqrt{1.1} = 1.04881$, and

$$1 + \frac{0.1}{2} - \frac{0.01}{8} + \frac{0.001}{16} - \frac{0.0005}{128} = 1.04881.$$

So by adding only the first five terms we obtain an estimate for $\sqrt{1.1}$ which is correct to five decimal places.

Application to $(1+1/n)^n$ We now apply the binomial theorem to the quantity $(1+1/n)^n$. The expansion we get on the right side is

$$1 + \frac{n}{1} \cdot \frac{1}{n} + \frac{n}{1} \cdot \frac{n-1}{2} \cdot \frac{1}{n^2} + \cdots$$

$$= 1 + \frac{1}{1} + \frac{1}{1} \cdot \frac{1-1/n}{2} + \frac{1}{1} \cdot \frac{1-1/n}{2} \cdot \frac{1-2/n}{3} + \cdots.$$

When n becomes indefinitely large, quantities such as $1/n$, $2/n$, ... become negligible, so the expression becomes

$$1 + \frac{1}{1} + \frac{1}{1 \times 2} + \frac{1}{1 \times 2 \times 3} + \cdots = 1 + \frac{1}{1!} + \frac{1}{2!} + \frac{1}{3!} + \cdots.$$

But when n increases without limit, the quantity $(1+1/n)^n$ gets closer and closer to e (this is the definition of e). It follows that

$$e = 1 + \frac{1}{1!} + \frac{1}{2!} + \frac{1}{3!} + \frac{1}{4!} + \cdots.$$

The proof that for all real numbers x, we have

$$e^x = 1 + \frac{x}{1!} + \frac{x^2}{2!} + \frac{x^3}{3!} + \frac{x^4}{4!} + \cdots,$$

follows in exactly the same way. This equation gives what is known as the *power series* for the exponential function.

As a corollary, we obtain

$$\frac{1}{e} = \frac{1}{2!} - \frac{1}{3!} + \frac{1}{4!} - \frac{1}{5!} + \cdots.$$

Power series for cosh and sinh The series for the cosh and sinh functions may now be obtained via the definitions ($\cosh x = (e^x + e^{-x})/2$, $\sinh x = (e^x - e^{-x})/2$); we get

$$\cosh x = 1 + \frac{x^2}{2!} + \frac{x^4}{4!} + \frac{x^6}{6!} + \cdots,$$

$$\sinh x = x + \frac{x^3}{3!} + \frac{x^5}{5!} + \frac{x^7}{7!} + \cdots.$$

Appendix C

Solutions

1.1.1 For 23×43 the HALVING sequence is 23, 11, 5, 2, 1; and the DOUBLING sequence is 43, 86, 172, 344, 688; so $23 \times 43 = 43 + 86 + 172 + 688 = 989$.

For 231×311 the HALVING sequence is 231, 115, 57, 28, 14, 7, 3, 1; and the DOUBLING sequence is 311, 622, 1244, 2488, 4976, 9952, 19904, 39808; so $231 \times 311 = 311 + 622 + 1244 + 9952 + 19904 + 39808 = 71841$.

1.1.2 The HALVING sequence is 16, 8, 4, 2, 1.

1.1.3 The HALVING sequence is 64, 32, 16, 8, 4, 2, 1; the only odd number in the sequence is 1, so in the doubling sequence the only number which contributes to the sum is the last one; i.e., $2^8 \times 103 = 64 \times 103$, which is just the product asked for.

1.1.4 Since $2^7 > 100$, in the HALVING sequence we reach a 1 in no more than 7 steps (all numbers between 64 and 99 will require 7 steps to reach a 1).

1.1.5 In the HALVING sequence only the very last entry is odd (it is a 1), so the only summand in the DOUBLING sequence is the last one, which is just the required product.

1.1.6 Examine what happens in a calculation such as 14×13. The HALVING sequence is 14, 7, 3, 1. The odd entries are at the 1st, 2nd and 3rd places from the right, corresponding to the fact that $14 = 2^1 + 2^2 + 2^3$. The DOUBLING sequence is 13, 26, 52, 104, and we must compute $26 + 52 + 104$, i.e., $(13 \times 2^1) + (13 \times 2^2) + (13 \times 2^3) = 13 \times (2^1 + 2^2 + 2^3) = 13 \times 14$. So the sum computed is just the required product. This reasoning is easily extended to the general case.

1.2.1 $1235 \times 571 = 705185$.

1.3.1 We have: $(319+211)/2 = 265$, $265^2 = 70225$; $(319-211)/2 = 54$, $54^2 = 2916$; and $70225 - 2916 = 67309$. So $319 \times 211 = 67309$.

Next, $(352+171)/2 = 261.5$, $261.5^2 = 68382.25$; $(352-171)/2 = 90.5$, $90.5^2 = 8190.25$; and $68382.25 - 8190.25 = 60192$. So $352 \times 171 = 60192$.

1.3.2 The size of the array would be 1000×1000; it would have one million entries. If the squaring idea is used, we need a table of squares from 1 till 2000 in steps of 0.5; we need four thousand entries. The saving is enormous.

1.4.1 $2117 + 22^2 = 51^2$, $\therefore$ $2117 = 51^2 - 22^2 = 29 \times 73$, and $2911 + 15^2 = 56^2$, $\therefore$ $2911 = 56^2 - 15^2 = 41 \times 71$.

1.4.2 $64777 + 48^2 = 259^2$, $\therefore$ $64777 = 211 \times 307$.

1.4.3 The method works well for numbers of the form ab, where a and b are of a similar magnitude. If n is of the form pq where p and q are primes and $p \ll q$ (i.e., p is much smaller than q), then the method will take a very long time to find the factors of n. Likewise, if n is prime the method will take a long time to show that it is prime.

2.1.1 1, 2, 4, 8, 16, 32, 64, 128, 256, 512, 1024.

2.1.2 1, 3, 9, 27, 81, 243, 729, 2187, 6561.

2.1.3 We have: $4^5 = 2^{10} > 2^9 = 8^3 > 7^3$, $\therefore$ $4^5 > 7^3$.

2.1.4 (a) $10^2 \times 10^3 = 10^5$ (b) $2^4 \times 2^5 = 2^9$ (c) $3^7 \div 3^4 = 3^3$ (d) $10^7 \div 10^3 = 10^4$ (e) $7^5 \div 7^2 = 7^3$.

2.1.5 $2^8 = 256$, $3^5 = 243$, $\therefore$ $2^8 > 3^5$.

2.1.6 Raising both sides of the relation $2^8 > 3^5$ to the 4th power, we get $2^{20} > 3^{12}$.

2.1.7 By computation, we get $3^7 = 2187$ and $2^{11} = 2048$; so 3^7 is the larger number.

2.1.8 2^{10}, 2^{12}.

2.1.9 $(3^4)^2 = 3^8$ and $(3^3)^4 = 3^{12} > 3^8$.

2.1.10 The numbers are 2^{30}, 2^{27} and 2^{32}, so the correct order is: 8^9, 2^{30}, 2^{32}.

2.1.11 They are equal to one another.

2.2.1 Taking one stride to be approximately 1 m, I run 10^4 m or 10 km; good enough for training purposes!

2.2.2 Taking the 'average' word to have 5 letters, the book has roughly 2×10^4 words; taking the average page to have 500 words, there are roughly $2 \times 10^4 \div 500$ or about 40 pages.

2.2.3 The thickness is $0.01 \times 2^{50} \approx 1.126 \times 10^{13}$ cm, i.e., it exceeds 10^8 km; this is about two-thirds of the Earth–Sun distance.

2.3.1 We have, $44^{44} = (44^{11})^4 > (44^2)^4 > 444^4$, and $4^{444} > 4^{440} = (4^{10})^{44} > 44^{44}$, as $4^{10} > 44$.

2.3.2 The answer is that $4^{4^{44}}$ is very much the larger number, as $4^{44} > 44^4$; for $4^{44} = (4^{11})^4 > 44^4$.

2.3.3 2^{222} has 67 digits (for $0.3 \times 222 = 66.6$).

2.3.4 Since 5^{10} is close to 10^7, we have $5 \approx 10^{0.7}$; so $5^{55} \approx 10^{0.7 \times 55} = 10^{38.5}$; so 5^{55} has roughly 39 digits. Next, $5^{555} \approx 10^{555 \times 0.7} = 10^{388.5}$, so it has roughly 389 digits. (In fact, the answer is 388.) Finally, $5^{5^{55}}$ has roughly 1.94×10^{38} digits.

2.3.5 The number is $1 + 2 + 2^2 + 2^3 + \cdots + 2^{24} = 2^{25} - 1 = 33554431$.

2.3.6 In a finite population, after a while, the persons spreading the rumour will communicate with people who already know the rumour; so the rate of growth will be substantially reduced.

2.4.1 2^{-2} and 10^{-6}.

2.4.2 $2^{20} \div 4^8 = 2^4$.

2.5.1 $16^{5/2} = 4^5 = 1024$, $36^{3/2} = 6^3 = 216$.

2.5.2 $27^{2/3} = 3^2 = 9$, $125^{5/3} = 5^5 = 3125$, $32^{6/5} = 2^6 = 64$.

2.5.3 $7^{(1+3+5+7)/4} = 7^4 = 2401$.

2.5.4 Expressing the fractions $1/2$ and $1/3$ with denominators of 6 (the LCM of 2 and 3), we get $2^{1/2} = (2^3)^{1/6} = 8^{1/6}$ and $3^{1/3} = (3^2)^{1/6} = 9^{1/6}$. Since $9 > 8$, we have $3^{1/3} > 2^{1/2}$.

2.5.5 $3^{1/4} = 27^{1/12}$ and $5^{1/6} = 25^{1/12}$, so $3^{1/4} > 5^{1/6}$.

2.5.6 (a) 10^3 (b) 2×10^2 (c) 3×10^3 (d) 4×10^2 (e) 14×10.

2.5.7 (a) 5 (b) 6 (c) 9.

2.5.8 (a) $4\frac{1}{2}$ (b) $5\frac{1}{2}$ (c) $19\frac{1}{2}$.

2.5.9 The least possible answer is $a = 2^3 \times 3^4 = 648$.

2.5.10 The least possible answer is $a = 2^{10} \times 3^6 \times 5^{15}$.

2.5.11 (a) $2^{1.5} \approx 2 \times 1.414 \approx 2.828$, $2^{2.5} \approx 5.626$ and $2^{3.5} \approx 11.312$.

(b) The answer is given in the text.

(c) If $2^x = 5$, then $2^{x+1} = 10$, so $x + 1 \approx 3.33$, so $x \approx 2.33$. (In fact, $2^{2.33}$ is roughly 5.02805.)

If $2^x = 9$, then $2^{2x} = 81 \approx 80 = 2^3 \times 10$, so $2^{2x-3} \approx 10$, so $2x - 3 \approx 3.33$, so $x \approx 3.165$. (In fact, $2^{3.165}$ is roughly 8.96933.)

2.5.12 $3^{0.5} \approx 1.732$, $3^{1.5} \approx 3 \times 1.732 \approx 5.196$, $3^{2.5} \approx 9 \times 1.732 \approx 15.588$.

2.5.13 (a) If $3^x = 10$, then $x \approx 2$. Better: since $3^4 = 81 \approx 2^3 \times 10$ and $2 \approx 10^{0.3}$, we get $3^4 \approx 10^{1.9}$, $\therefore 3 \approx 10^{0.475}$. So $x \approx 1/0.475$, i.e., $x \approx 2.1$. (In fact, $3^{2.1}$ is roughly 10.0451.)

(b) If $3^x = 20 = 2 \times 10$, then $3^x \approx 10^{1.3}$, so $x/1.3 \approx 2.1$, $\therefore x \approx 1.3 \times 2.1 = 2.73$. (In fact, $3^{2.73}$ is roughly 20.0697.)

(c) If $3^x = 50 = 100/2$, then $3^x \approx 10^2 \div 10^{0.3} = 10^{1.7}$, so $x/1.7 \approx 2.1$, $\therefore x \approx 1.7 \times 2.1 = 3.57$. (In fact, $3^{3.57}$ is roughly 50.5037.)

2.6.1 According to Kepler's III Law, T is proportional to $r^{3/2}$.

2.6.2 The ratio of volumes is $\left[(1.39 \times 10^5)/(1.22 \times 10^4)\right]^3$, or about 1480.

2.6.3 The ratio of the mass of Jupiter to the mass of the Solar System is $1.899 : 1993$ or roughly $1 : 1050$. The Sun contains more than 99.8% of the mass of the Solar System.

2.6.4 The number of H atoms in the Sun is roughly equal to the mass of the Sun divided by the mass of a proton, or about $(1.99 \times 10^{30})/(1.67 \times 10^{-27}) \approx 1.2 \times 10^{57}$; i.e., about 10^{57}.

2.6.5 The number of H atoms that would make up the Earth is about $(5.98 \times 10^{24})/(1.67 \times 10^{-27}) \approx 3.6 \times 10^{51}$.

2.6.6 The ratio is about $\dfrac{9.1 \times 10^{-28}}{1.67 \times 10^{-24}} \approx 1835$.

2.6.7 The number of seconds in a year is $60 \times 60 \times 24 \times 365 = 31536000$, or roughly 3.15×10^7.

2.6.8 As my age is 50 years, I am about $50 \times 3.15 \times 10^7 \approx 1.57 \times 10^9$ seconds old; i.e., my age is roughly $1\frac{1}{2}$ billion seconds. How about you?

2.6.9 The number of beats is roughly $72 \times 60 \times 24 \times 365 \times 75 \approx 2 \times 10^9$; i.e., about 2 billion.

2.6.10 The age of the universe is roughly $2 \times 10^{10} \times 3.15 \times 10^7$ or about 6.3×10^{17} seconds.

2.6.11 A light year is $3 \times 10^5 \times 3.15 \times 10^7 \approx 9.45 \times 10^{12}$ km; i.e., roughly 10^{13} km.

2.6.12 The Sun-to-Earth distance is $(14.86 \times 10^7)/(3 \times 10^5)$ or about 495 light seconds.

The Sun-to-Pluto distance is $(587.2 \times 10^7)/(3 \times 10^5) \approx 19573$ light seconds. This is about 326 light minutes; i.e., about $5\frac{1}{2}$ light hours.

3.1.1 Both sides equal 2047.

3.1.2 Divide both sides by 2^n.

3.1.3 Let $1 - 1/2 + 1/4 - 1/8 + \cdots \pm 1/2^n = A$. (The sign of the last term is negative if n is odd, positive if n is even.) Dividing each term by 2 we get $1/2 - 1/4 + 1/8 - \cdots \pm 1/2^{n+1} = A/2$. When we add the two equations and 'cancel' all negative-positive pairs, we get $3A/2 = 1 \pm 1/2^{n+1}$. As n gets larger, the fraction $1/2^{n+1}$ gets smaller, and 'in the limit' it vanishes; so A tends to $2/3$.

3.2.1 Both sides equal 29524.

3.2.2 Let $1 + 1/3 + 1/9 + \cdots = B$. Dividing by 3, we get $1/3 + 1/9 + 1/27 + \cdots = B/3$. Subtraction yields $1 = 2B/3$, so $B = 3/2$.

3.2.3 Let $1 - 1/3 + 1/9 - 1/27 + \cdots = C$. Dividing by 3, we get $1/3 - 1/9 + 1/27 - 1/81 + \cdots = C/3$. Addition yields $1 = 4C/3$, $\therefore C = 3/4$.

3.2.4 The sum is $(10^n - 1)/9$.

3.2.5 The sum $1 + 1/10 + 1/10^2 + \cdots + 1/10^{n-1}$ is equal to $(1 - 1/10^n)/(1 - 1/10)$. As n gets larger and larger, the term $1/10^n$ shrinks to 0, and we are left with the quantity $10/9$.

3.2.6 Let $1 - 1/10 + 1/100 - 1/1000 + \cdots = D$. Dividing by 10, we get $1/10 - 1/100 + 1/1000 - 1/10000 + \cdots = D/10$. Addition yields $1 = 11D/10$, $\therefore D = 10/11$.

3.2.7 The sum is $(a^n - 1)/(a - 1)$.

3.2.8 The sum is $a/(a - 1)$.

3.3.1 $1/13 = 0.\overline{076923}$, $1/17 = 0.\overline{0588235294117647}$.

3.3.2 The decimal $0.1111\ldots$ is the same as $1/10 + 1/100 + 1/1000 + 1/10000 + \cdots$.

3.3.3 The decimal $0.01010101\ldots$ is the same as $1/10^2 + 1/10^4 + 1/10^6 + \cdots$.

3.3.4 Since $1/100 + 1/100^2 + 1/100^3 + \cdots = 1/99$, we get $a/b = 18/99 = 2/11$.

3.3.5 Here $1/b + 1/b^2 + 1/b^3 + \cdots = 1/999999$, so

$$\frac{a}{b} + \frac{a}{b^2} + \frac{a}{b^3} + \cdots = \frac{142857}{999999} = \frac{1}{7}.$$

3.3.6 Since $1/1000 + 1/1000^2 + 1/1000^3 + \cdots = 1/999$, we get $0.123\,123\,123\ldots = 123/999 = 41/333$.

3.4.1 The sum of the series is $11.1111\ldots = 100/9$.

3.4.2 The paradox is resolved when we observe that $1/2 + 1/4 + 1/8 + \cdots = 1$.

3.4.3 As we go back in time, our ancestors are our ancestors in many different ways. In other words, the ancestors are related to one another (cousins marrying one another, cousins marrying uncles, and so on).

3.4.4 On the first leg of the journey, the bee and A meet after $100/30$ or $10/3$ hours, so the bee travels $200/3$ km. In this time, B has travelled $100/3$ km. On the second leg of the journey, the bee and B meet after $200/3 \div 30 = 20/9$ hours, so the bee travels $200/9$ km. Similarly, on the 3rd leg, the bee travels $200/29$ km, and so on. So the total distance travelled by the bee is $200/3 + 200/9 + 200/27 + \cdots$, i.e., $200(1/3 + 1/9 + 1/27 + \cdots) = 200 \times 1/2 = 100$ km. So the bee travels 100 km in all.

There is a shorter way of arriving at this answer! Observe that A and B meet after $100/20 = 5$ hours, and during these 5 hours the bee is continually moving between the two cyclists. So it travels $5 \times 20 = 100$ km.

4.1.1 $\log_4 16 = 2$, $\log_5 125 = 3$.

4.1.2 $\log_9 3 = 0.5$, $\log_{32} 16 = 4/5 = 0.8$.

4.1.3 The sum is $\log_5(1 \times 3 \times 5 \times 7 \times 9) = \log_5 945$.

4.1.4 Let $\log_a b = x$; then $a^x = b$, $\therefore a = b^{1/x}$, $\therefore \log_b a = 1/x$.

4.1.5 Let $\log_a b = x$; then $a^x = b$, $\therefore (1/a)^x = 1/b$, $\therefore \log_{1/a} 1/b = x$. The two quantities are equal.

4.1.6 Let $\log_a b = x$, $\log_b c = y$, $\log_c a = z$; then $a^x = b$, $b^y = c$, $c^z = a$, $\therefore a^{xyz} = 1$, $\therefore xyz = 1$, $\therefore 1/y = xz = x/(1/z)$, i.e., $\log_c b = \log_a b / \log_a c$.

4.1.7 $\log_a b = n/m$.

4.1.8 $\log A = \log \pi + 2\log r$, $\log V = \log 4/3 + \log \pi + 3\log r$ (all logarithms to the same base).

4.1.9 We must show that $\log a^{1/n} = (1/n) \times \log a$; but this is obvious.

4.1.10 The logarithm of $a^{\log b}$ is $\log b \log a$, and the logarithm of $b^{\log a}$ is $\log a \log b$. As the two quantities have the same logarithm, they are equal.

4.4.1 We have, $3^3 = 27$, $5^2 = 25$. Taking 25 and 27 to be 'close' to one another, we get $3^3 \approx 5^2$, or $3^{3/2} \approx 5$. So $\log_3 5 \approx 1.5$. (The actual value is 1.46497.)

Next, $5^3 = 125$ and $11^2 = 121$, and these are 'close' to one another; so $5^3 \approx 11^2$, $5^{3/2} \approx 11$, $\log_5 11 \approx 1.5$. (The actual value is 1.4899.)

4.4.2 They are equal!

4.4.3 Observe that $(2/3)\log_2 3 = \log_8 9$ which exceeds 1, whereas $(2/3)\log_5 10 = \log_{125} 100$ which is below 1. Therefore, $\log_2 3 > \log_5 10$.

4.4.4 We shall show that the fraction $7/4$ lies between the two given quantities. Since $3^7 = 2187$ and $7^4 = 2401$, we get $3^7 < 7^4$, $\therefore 3^{7/4} < 7$, $\therefore \log_3 7 > 7/4$. Next, $5^7 = 78125$ and $16^4 = 2^{16} = 65536$, $\therefore 5^7 > 16^4$, $\therefore \log_5 16 < 7/4$. So $\log_3 7 > \log_5 16$.

4.6.1 $\log_{10} 12 = (2 \times 0.3010) + (0.4771) = 1.0791$; $\log_{10} 120 = 2.0791$. (In fact, the answers are 1.0792 and 2.0792, respectively.

Loss of accuracy has occurred because of the round-off.) Next, $\log_{10} 576 = \log_{10}(2^6 \times 3^2) = (6 \times \log_{10} 2) + (2 \times \log_{10} 3) = 2.7602$. (The actual value is 2.7604.) Finally, $\log_{10} 48 = \log_{10}(2^4 \times 3) = (4 \times \log_{10} 2) + \log_{10} 3 = 1.6811$. (The actual value is 1.6812.)

4.6.3

$$\log_{16} 24 = \frac{\log_2 24}{\log_2 16} = \frac{\log_2(2^3 \times 3)}{4}$$

$$= \frac{3\log_2 2 + \log_2 3}{4} = \frac{3 + c}{4},$$

$$\log_{12} 24 = \frac{\log_2 24}{\log_2 12}$$

$$= \frac{3\log_2 2 + \log_2 3}{2\log_2 2 + \log_2 3} = \frac{3 + c}{2 + c},$$

$$\log_{18} 36 = \frac{\log_2 36}{\log_2 18} = \frac{\log_2(2^2 \times 3^2)}{\log_2(2 \times 3^2)}$$

$$= \frac{2\log_2 2 + 2\log_2 3}{\log_2 2 + 2\log_2 3} = \frac{2 + 2c}{1 + 2c}.$$

4.6.4 Let $\log_2 5 = b$; then

$$\log_{20} 80 = \frac{\log_2 80}{\log_2 20} = \frac{\log_2(2^4 \times 5)}{\log_2(2^2 \times 5)}$$

$$= \frac{4 + b}{2 + b}.$$

So $a = (4 + b)/(2 + b)$, $\therefore 2a + ab = 4 + b$, $\therefore (a - 1)b = 4 - 2a$, so $b = (4 - 2a)/(a - 1)$. Similarly,

$$\log_{10} 50 = \frac{\log_2 50}{\log_2 10} = \frac{\log_2(2 \times 5^2)}{\log_2(2 \times 5)}$$

$$= \frac{1 + 2b}{1 + b} = \frac{1 + 2(4 - 2a)/(1 - a)}{1 + (4 - 2a)/(1 - a)}$$

$$= \frac{9 - 5a}{5 - 3a}.$$

5.1.1 We get

$$\log_{10} 5 = 0.10110010111011\ldots,$$
$$\log_{10} 7 = 0.1101100001011\ldots.$$

5.1.2 Let $\log_{10} x = 0.a_1 a_2 a_3 a_4 \ldots$ (in binary form). If $a_1 = 0$, then $\log_{10} x \leq 0.0111 \ldots < 0.1 = 1/2$, so $x \leq \sqrt{10}$. If $a_1 = 1$, then $\log_{10} x \geq 0.1$, so $x \geq \sqrt{10}$.

Next, $0.a_2 a_3 a_4 \ldots = \log_{10} x^2$ or $\log_{10}(x^2/10)$, depending on whether $x^2 < 10$ or $x^2 > 10$. So if $x^2 < 10$, then $a_2 = 0$ if $x^4 < 10$ and $a_2 = 1$ if $x^4 \geq 10$. And if $x^2 \geq 10$, then $a_2 = 0$ if $(x^2/10)^2 < 10$ and $a_2 = 1$ if $(x^2/10)^2 \geq 10$.

5.1.3 This should be clear from **5.1.2**.

5.2.1 The details are left to the reader; we get $\log_{10} 3 = 0.4771$ and $\log_{10} 7 = 0.8451$ (roughly).

5.2.2 We divide or multiply x by a suitable integral power of 10 so that the resulting number lies between 1 and 10, and then we work with this number rather than the original one.

5.3.1 We have: $10^{3/8} = 10^{1/4} \times 10^{1/8} = 1.77828 \times 1.33352 = 2.37137$.

5.3.2 $10^{5/16} = 2.05353$ and $10^{5/32} = 1.43301$.

5.3.3 $\log_{10} 6 \approx 0.778151$.

5.5.1 $\log_{10} 11 = 1.04139$ and $\log_{10} 13 = 1.11394$.

5.5.2 We shall use the relation

$$\ln \frac{1+h}{1-h} \approx 2\left(h + \frac{h^3}{3} + \frac{h^5}{5}\right).$$

Let $h = 1/3$; though this is certainly not small in comparison with 1, there is no harm in seeing what happens. We get

$$\ln 2 \approx 2\left(\frac{1}{3} + \frac{1}{3 \times 3^3} + \frac{1}{5 \times 3^5}\right) = 2\left(\frac{1}{3} + \frac{1}{81} + \frac{1}{1215}\right)$$

$$= 2(0.333333 + 0.0123457 + 0.000823045) = 0.693004.$$

So $\ln 2 \approx 0.693$. The inclusion of more terms in the power series expansion for $\ln \dfrac{1+h}{1-h}$ will yield more accurate values.

6.2.1 Since $\ln 3 \approx 1.1$, the number of years needed for the money to triple with an interest rate of $r\%$, is roughly $110/r$.

6.2.2 There are numerous factors that complicate the population growth profile; war, natural disasters, deaths due to famine, and so on. We have, as well, the effect observed in the industrialized countries of a steady decrease in population growth rate with an increasing level of affluence.

9.1.1 For two stars A and B possessing equal absolute magnitude, we have

$$m_A - m_B = 2.512 \log_{10} \frac{l_B}{l_A}.$$

Since the stars have equal absolute magnitude, the light flux is inversely proportional to the square of the distance; so $l_B/l_A = (r_A/r_B)^2$. Therefore, by substitution, we get

$$m_A - m_B = 2.512 \log_{10} \left(\frac{r_A}{r_B} \right)^2 = 5.024 \log_{10} \frac{r_A}{r_B}.$$

Approximating 5.024 by 5, we get $m_A - m_B \approx 5 \log_{10} r_A/r_B$. This equation yields the desired result (think of A and B as being the same star, placed respectively at distances of r light years and 10 light years from the Earth).

10.4.1, 10.4.2, 10.4.3 The limits are $\sqrt{e}$, $e^{1/3}$ and e^2, respectively.

10.4.4 The limit for any fixed value of x of $(1 + x/n)^n$, as n gets larger and larger, is e^x.

11.8.1 The sequence $\{u_n\}$ steadily increases, whereas $\{v_n\}$ steadily decreases. Both have e as their limit.

11.8.2 The proof is given in Appendix A.

13.5.1 We have: $L(t^2) = L(t \times t) = L(t) + L(t) = 2L(t)$, $L(t^3) = L(t^2 \times t) = L(t^2) + L(t) = 3L(t)$, and so on.

13.5.2 This is true for any shape. A contraction along the x-axis by a factor of 2 reduces the area by half; and then an expansion along the y-axis by a factor of 2 doubles the area. So the net effect is—no change in area.

Bibliography

[Bea]   Beatty, Samuel. *Elementary Proof That e is not Quadratically Algebraic*, in *A Century of Calculus, Part I*. Mathematical Association of America, 1969.

[Bel]   Bell, Eric Temple. *Men of Mathematics*. Simon and Schuster, 1937.

[Caj]   Cajori, Florian. *A History of Mathematics*. Macmillan, 1919.

[Coo]   Coolidge, Julian L. *The Number e*, in *A Century of Calculus, Part I*. Mathematical Association of America, 1969.

[Eve1]   Eves, Howard. *Great Moments in Mathematics I*. Mathematical Association of America, 1983.

[Eve2]   Eves, Howard. *Great Moments in Mathematics II*. Mathematical Association of America, 1983.

[Kli]   Kline, Morris. *Mathematical Thought from Ancient to Modern Times*. Oxford University Press, 1972.

[Kra]   Kramer, Edna E. *The Nature and Growth of Modern Mathematics*. Princeton University Press, 1981.

[Mac]   Macleish, John. *Number*. Flamingo, 1992.

[Mao1]   Maor, Eli. *e: The Story of a Number*. Universities Press, 2000.

[Mao2]   Maor, Eli. *Trigonometric Delights*. Universities Press, 2000.

[Sim1]   Simmons, George F. *Calculus with Analytic Geometry.* McGraw-Hill, 1985.

[Sim2]   Simmons, George F. *Differential Equations with Applications and Historical Notes.* McGraw-Hill, 1972.

[Tho]   Thompson, D'Arcy W. *On Growth and Form.* Cambridge University Press, 1917.

Numerous readings have been listed above. Of these, the following are particularly recommended:
— [Coo], for a very readable historical account of e;
— [Eve1] and [Eve2], for highly readable and informative accounts of various historically significant moments ("great moments") in the history of mathematics, including the history of logarithms;
— [Mao1], for a wonderful fund of information on virtually every aspect concerning e; and
— [Sim2], for rigorous accounts of many problems where e enters naturally; e.g., the problem of the hanging chain.

From the point of view of general reading, however, *all* the readings are strongly recommended.

Index

187